The Power of the DUCHESSES

Duchess Power on Grayrigg

Eric Treacy

Frontispiece: Duchess Power – streamlined style
Eric Treacy

The Power of the DUCHESSES

Compiled by David Jenkinson

Foreword by R. A. Riddles, CBE

© Oxford Publishing Co and D. Jenkinson

ISBN 978 1 909625 16 7

Jacket photographs
City of Nottingham, resplendent in BR Maroon livery, from transparencies by Tim Shuttleworth.

This book is dedicated to the memory of
GAVIN LAURIE WILSON
. . . one of my very dearest friends and one of nature's gentlemen. His love for the Duchesses knew no bounds and kindled my own fascination with Stanier's wonderful engines — while his files and records have played a major part in the compilation of this book. Thanks for the memory, Gavin, I hope you would have approved.

Coronation at 114 mph — R. A. Riddles on the footplate.

BR LMR

Printed by The Amadeus Press, Cleckheaton, BD19 4TQ

Published by
Book Law Publications
382 Carlton Hill
Nottingham
NG4 1JA

FOREWORD

How fortunate it is to be in the right place at the right time! Shortly before the arrival of Sir William Stanier as Chief Mechanical Engineer to the LMS Railway, I had been moved back to Crewe from Derby. At this time, Stanier's appointment was a complete cure to the internal and eternal friction as between Derby, Crewe, Horwich and St Rollox — and I was there to see the new dawn breaking!

Stanier established himself quickly and soon the first of his Pacifics was under way — and without these pioneer engines there would have been no Duchesses to follow them a few years later. I became very much involved in the building of the first 4-6-2 (No. 6200 *Princess Royal*). New ideas and new techniques had to be adopted and such was the urgency and pressure that the engine was only just completed in time to make its first trip outside the works direct to Euston for Official inspection. With this class, the stage was set for future plans to be developed and some time later, Stanier sent for me and told me I was to come to London as his Principal Assistant — no ifs and buts — you just accepted the decision.

Thus it was that I became particularly involved with the evolution of the Duchesses and we discussed at our first meeting the future of the 4-6-2 type. Increased speeds and faster timings were being asked for so, remembering some advice I had been given that there was an optimum piston speed after which efficiency was lost, I suggested an increase in driving wheel diameter and this was accepted. In retrospect I wish it hadn't for I later became much more concerned with adhesion and acceleration — hence my own later work with 2-10-0s. Pacifics tend to be bad starters and require careful handling — the Duchesses being no exception — and I don't believe they would have been any the worse with somewhat smaller wheels. Nevertheless, they were magnificent machines and a considerable advance over the first LMS 4-6-2s.

Sir William was a marvellous chief and I owe so much to him that I am delighted to have an opportunity to introduce this very excellent record which Mr Jenkinson has compiled of the Duchesses. Although much has been written and said about them from time to time, this very complete historical survey of the engines in all their guises, demonstrates very fully not only the success of the locomotives themselves but places on record once again a great achievement by a very very eminent locomotive engineer.

R.A. Riddles, CBE
Calne 1978

Streamlined *Princess Alice* heads the down 'Coronation Scot' at Northchurch in 1938.
E. R. Wethersett

THE POWER OF THE DUCHESSES

... Midsummer dawn and the chorus of birdsong is just beginning to disturb the silence of the short summer night over the Cumbrian fells when, away to the south at Tebay, the sharpening four beat exhaust of a large steam locomotive working hard can just be discerned above the timeless sounds of nature in this desolate yet magnificent spot. Soon a plume of steam and smoke can be seen erupting skywards from the cutting below Scout Green and into view comes a gleaming dark red Pacific locomotive. Behind it, . . . one, two, three, . . . twelve, thirteen, fourteen (or even more) massive sleeping cars of the 'Night Scot' en route to Glasgow — perhaps 600 tons of train proceeding majestically up one of Britain's most famous railway gradients behind what is arguably the supreme example of express passenger steam power ever to be built in Britain — an LMS 'Duchess' 4-6-2.

If ever a British steam locomotive deserved to be commemorated as the epitome of power then it must be Stanier's immortal design — but such a statement is suspect without qualification. The railway historian must strive to be impartial and the compiler of a book must beware of letting his prejudices stand in the way of truth. Nevertheless, this writer, when asked by the publisher to compile a book on the *Power of the Duchesses* was in no real doubt that of all the steam locomotives built for use in Britain, there could be no more worthy contender, no matter what yardstick is applied, than the final LMS design of 4-6-2.

Let us examine the claim. What is power? I seem to remember, dimly from my schooldays, that it is something to do with the 'capacity to perform work'. Now work is something actually achieved — not just that which is theoretically available as a result of some abstruse calculation of basic dimensions. On the latter score there were several British locomotives which could claim, on the basis of nominal tractive effort, to be more 'powerful' than the Duchesses. But it is at the cylinders and drawbar where the work is actually done and on this basis, Stanier's great engines were never surpassed in this country — approached, yes; but never outperformed. Even before the war, a Duchess had indicated over 3300hp at the cylinders and this sort of performance potential was regularly approached both in traffic and on test. Even in the matter of speed, the Duchesses were only bested by Gresley's equally celebrated A4 type — and then only on one or two occasions. Yes, I am prejudiced in favour of the LMS design but not, I hope, without good authenticated evidence to back it up.

This is not, however, a technical treatise — that side of the story is in print elsewhere for those who would seek it out. Rather, it is a pictorial celebration, at times quite sentimental, of a locomotive type which even those whose favourites ran on other routes will readily admit was a noteworthy design, given the constraints on absolute steam power imposed by the British loading gauge.

I am told that this is the first full length pictorial survey wholly devoted to the Stanier Duchesses and for this reason, if no other, I feel that it is a little impertinent of me to be compiling it since most of the credit must, rightly, belong to the photographers. The best I can hope to do is arrange their work in such a way as to tell the story in appropriate manner — and I am more grateful than words can tell for the help I have received from so many well known exponents of the difficult art of railway photography. And I am sure none of the gentlemen concerned would object if I singled out for especial mention the work of my friend the late Bishop Eric Treacy. Shortly before he died, he allowed me to choose the very pick of his collection and I am deeply grateful to have this chance to publish so much of his work as my own personal 'thank you'.

Some of the pictures in this book have appeared before — usually in books or articles long out of print — but I make no apology for their re-use; for they are in the nature of classic views without which I feel the survey would be incomplete. There is, however, much that is new (both from private and official sources) and I can only hope that those who were even more familiar with the Duchesses than myself will take pleasure in the chosen 'mix'. As for those who never knew the engines in their pomp and prime, I hope this book will be some consolation for what they missed.

Finally, I am greatly indebted to Robin Riddles for having agreed to write a foreword, for he is a supremely modest man who never seeks the limelight. Mr Riddles was William Stanier's Principal Assistant in the 1930s, was on the footplate of *Coronation* herself on that epic run down Madeley bank in 1937 and took 6229 (disguised as 6220) to North America in 1939. If any man knows the virtues of these engines then it must surely be he. I count myself privileged to have met him in recent years for he surely merits the title of Britain's 'Elder Statesman' of steam locomotive engineering — an accolade he inherited, fittingly, in succession to his old chief, Sir William Stanier.

And now . . . let the Duchesses speak for themselves. . . .

D. Jenkinson
Knaresborough 1979

EVOLUTION AND BUILDING

Of course, they weren't always called 'Duchesses' and, indeed, only ten of the 38 examples built were ever named after these exalted ladies. The official designation was 'Princess Coronation', later shortened to 'Coronation' class and it was only after the first five appeared with Royal names (by permission of the Palace) that the 'Duchess' theme was established. In spite of a later proliferation of the 'City of . . .' theme, the Duchess names quickly captured popular imagination — for they were surely 'Duchesses' of the rails — dignified, impressive and majestic.

The class was derived from the earlier Stanier 'Princess Royal' design of 1933-5, the detail work being largely the result of efforts by Stanier's chief draughtsman Tom Coleman, who felt that a better engine could be achieved than by simply building more 'Princesses' — which was the original intention for the 1937 high speed services. That this is so is evidenced by recent material found in the archives of the National Railway Museum and reproduced here for the first time as far as is known. A diagram was proposed for further modifications to the earlier 4-6-2 type and the scheming even went as far as a proposed streamline casing.

There was even some toying with the idea of a water tube firebox. However, Coleman persuaded Stanier to agree to a slightly less radical re-design, which nevertheless took matters much further away from the classical Swindon/Churchward ideas than once seemed likely; and the new engines were actually designed in detail during Stanier's absence in India. This was really quite a tremendous tribute to the way Stanier had transformed the LMS — for he surely knew that it would be he who would 'carry the can' were the new design to fail, regardless of who actually designed what.

Essentially, the Duchesses differed from the Princesses in having a much better (and larger) boiler, slightly larger driving wheels (for higher speed), a much simplified valve gear (two sets instead of four) and a vastly improved 'front end'. They were, of course, given streamline casings to suit the fashion of the day. On the next few pages, the evolution of the first examples is illustrated from the preliminary design studies for an improved Princess through to the final 'rolling out' of the first three engines.

Plate 1 (left): Stanier Princess Royal Class No. 6203 *Princess Margaret Rose* — the design from which the Duchess type was derived.
NRM

Plate 2 (below): Unfulfilled proposal for a modified 6ft 6in wheeled 4-6-2 with closer coupled wheelbase and altered front end — clearly part way to the Duchess concept.
NRM

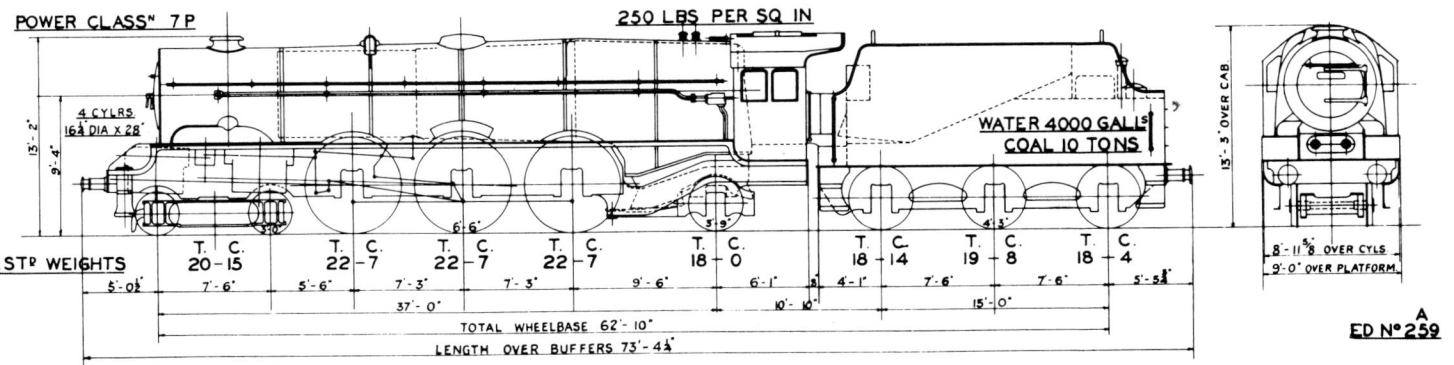

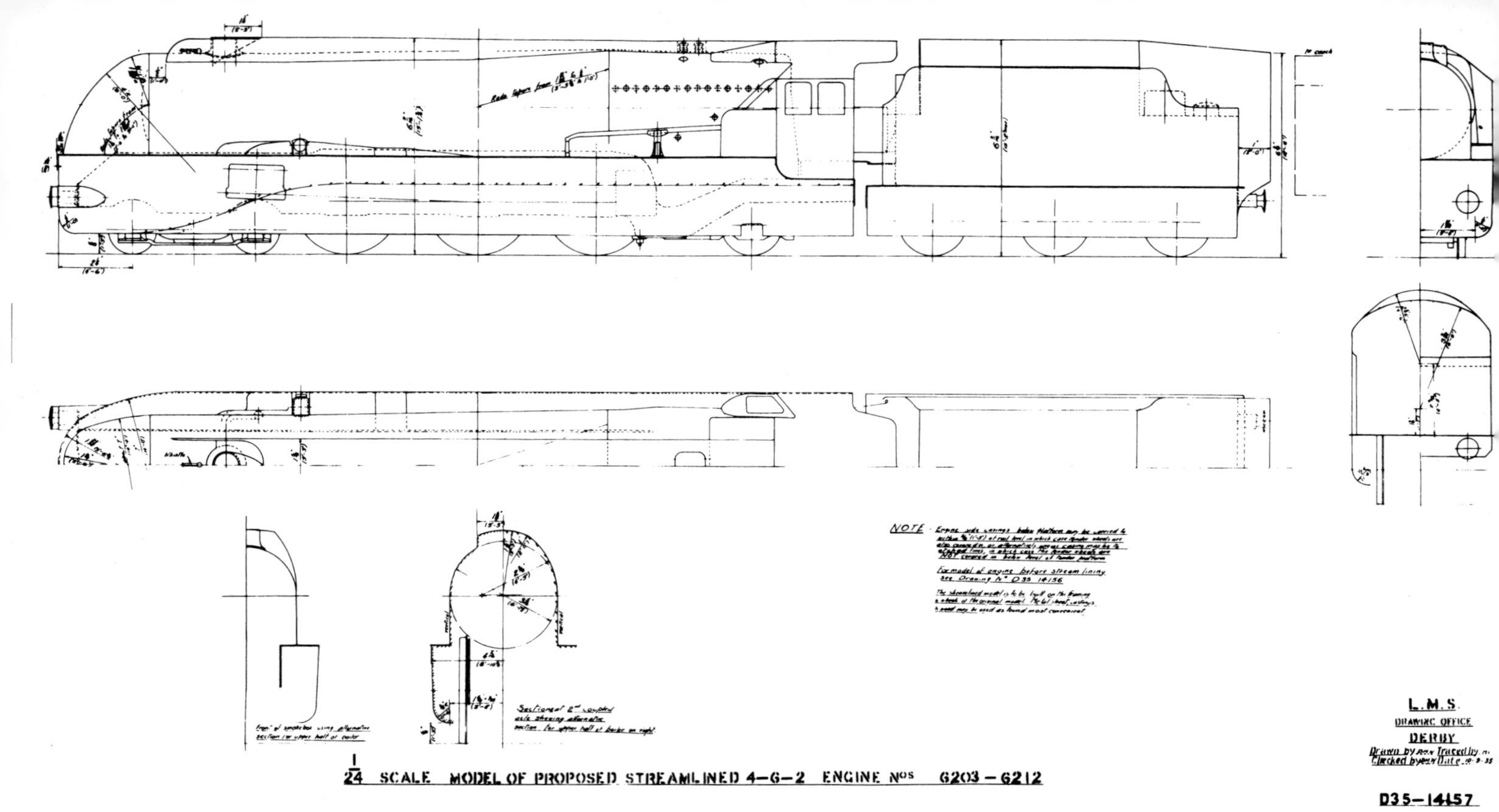

Plate 3-8: On this page (*above*) are shown diagram and 1/24 scale model of a proposed streamlined version of the Princess Royal type, dated 1935. The influence of contemporary German styling is evident. *Below* is shown a later streamlined model (basically the shape finally adopted) which is believed to have been that which was wind tunnel tested prior to the building of the Duchesses. *On the opposite page*, the pioneer streamlined engine (No. 6220) is seen at various stages of erection at Crewe.

NRM

Plate 9-12: Frames, firebox backhead, bogie and trailing truck of No. 6220 under construction at Crewe.

NRM

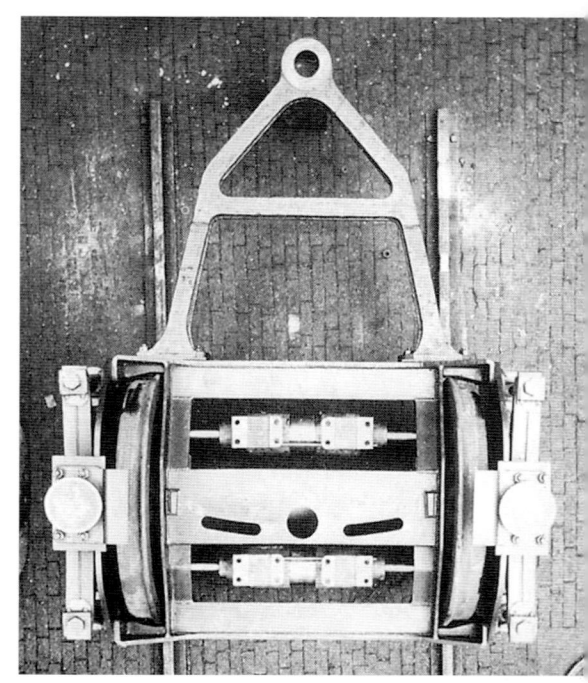

Plate 13-15: Above can be seen the cylinders and smokebox (with single chimney) of No. 6220 while *below* is a view of the pioneer engine almost complete, showing the method of attachment of the streamline casings.

NRM

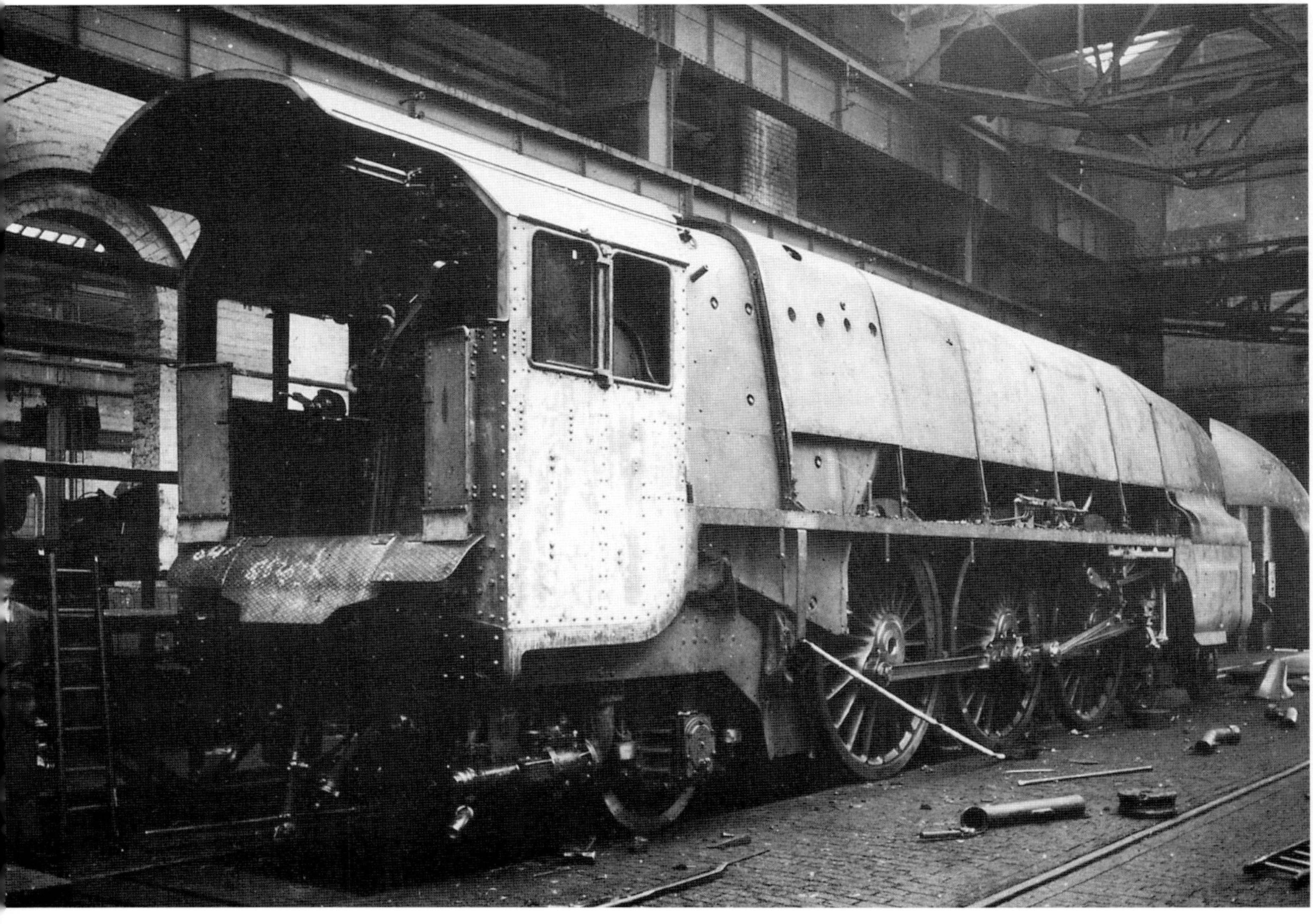

Plate 16-19: Coronation, Queen Elizabeth and *Queen Mary*, the first three blue streamliners, were lined up proudly for inspection at Crewe. With their dark blue wheels and nameplates and polished motion and trimmings, they must have made an impressive sight. *Below* is a more detailed view of *Queen Mary*. Incidentally, the nameplates on these engines had their letters and surrounds chrome plated in addition to the special blue background colour.

NRM

THE BLUE STREAMLINERS

The first five engines, painted in 'Caledonian' blue with silver stripes (at Mr. Riddles' suggestion), emerged in 1937 and were immediately put to work on the new 'Coronation Scot' prestige express. On her press run with the train, No. 6220 immediately set a new steam record of 114mph (and still accelerating) when safety considerations caused a brake application. No one will ever know how fast she could have travelled for the LMS had no stretch of line on which, at that time, higher speed could have been attained. No matter, 'Coronation's' exploits undoubtedly spurred the East Coast to go even faster and thus directly led to the 126mph achieved by *Mallard* the following year.

Plate 20: This view of No. 6223 *Princess Alice* at Edge Hill shows one of the blue streamliners engaged on other than 'Coronation Scot' duties — this time backing down to Liverpool Lime Street for a West of England train.

Eric Treacy

Plate 21,22: Of course, the blue streamliners will for ever be associated with the 'Coronation Scot'. These two Eric Treacy views show the first and the last of them at work in the north of England. *Above* is No. 6224 *Princess Alexandra* northbound near Preston while *to the left Coronation* herself heads south in the Weaver Junction area.

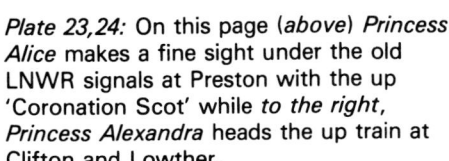

Plate 23,24: On this page (*above*) *Princess Alice* makes a fine sight under the old LNWR signals at Preston with the up 'Coronation Scot' while *to the right*, *Princess Alexandra* heads the up train at Clifton and Lowther.

Eric Treacy, BR LMR

THE RED STREAMLINERS

After the first five blue streamliners, the LMS, keeping the same lining style (now, however, rendered in gold) reverted to the familiar Midland Red colour and built fifteen more locomotives (Nos. 6225-9, 35-44) in the new livery. The missing numbers (Nos. 6230-4) were given to non-streamlined examples and we will meet these machines later. The second world war was fast approaching and few of the later streamliners were photographed in red livery as much as the blue ones had been. Of course, they mostly went into general service, the blue ones being preferred, where possible, for the 'Coronation Scot' train. In time, however, the smart turnout deteriorated and they became filthier and filthier as the war progressed.

Plate 25,26: The first red streamliner, No. 6225 *Duchess of Gloucester* also established the 'Duchess' theme. Two head on views are given on this page. *Above*, No. 6225 is seen being serviced at Perth in August 1939 while *below* she is seen leaving Carlisle with the down 'Royal Scot' — if truth be told, a somewhat more demanding job than the more glamorous 'Coronation Scot'.

Gavin Wilson

Plate 27,28: The much more varied tasks given to the later streamliners are well exemplified on this page. *Above,* No. 6228 *Duchess of Rutland* heads past Bushey troughs with the down 'Royal Scot' and in the process gives something of the lie to the story that the LMS streamline casing was not very good at keeping smoke out of the driver's view. *Below, Duchess of Gloucester* is seen at Edge Hill in front of one of those so very typical LMS cavalcades of assorted designs of coaches — seemingly stretching back for ever from the engine.

Real Photographs, Eric Treacy

Plate 29: The second batch of red streamliners emerged after the decision to adopt the double chimney arrangement. This somewhat unusual view at Crewe shows the new feature very clearly on No. 6236 *City of Bradford*.
Real Photographs

Plate 30, 31 (below): The LMS got itself into something of a muddle with its naming of the red streamliners. No. 6244 did indeed start life as *City of Leeds* but quickly became *King George VI* in 1941; but No. 6253 was never *City of Sheffield*, nor was it streamlined for that matter! The explanation is that the first double chimney streamliner No. 6235 was photographed successively in works grey livery carrying the number and name of all the then planned streamliners before the engines actually emerged. Subsequent wartime events were to make a nonsense of some of the official pictures taken at the time. Both these views are of the same engine!
NRM

THE TRANSATLANTIC WANDERER

One of the red streamliners (No. 6229 *Duchess of Hamilton*) had a much more exciting debut than did her fourteen similarly garbed sisters, for the LMS had accepted an invitation to exhibit its proposed new 'Coronation Scot' train at the New York World's fair in 1939 and it was decided to send one of the new batch of engines with the train. No. 6229 was, therefore, put to work in shop grey finish for running-in purposes and then re-entered works to re-emerge resplendent in Midland Lake but carrying the number and name No. 6220 *Coronation*. The real No. 6220, in the meantime, was renumbered No. 6229 and became *Duchess of Hamilton* for a few years — still painted blue, however. The red one was fitted with typical American pattern headlight and bell and, with her train, proceeded to captivate the USA until her return in 1942 when identities were exchanged again. Fortunately, these activities were well recorded and the next few pages recount the story.

Plate 32-4: *Duchess of Hamilton* as herself, just after building. *At the right*, she is seen posed in primer paint at Crewe alongside the famous *Lion*, built 100 years earlier for the Liverpool and Manchester Railway. *Below, left*, the engine is seen posed for the official picture in shop grey livery while *at the right*, still in shop grey, No. 6229 was photographed on the standard Crewe running-in turn at Shrewsbury.

NRM, Real Photographs

Plate 35: Fireman J. Carswell and Driver F.C. Bishop on the footplate of No. 6220, now in red livery for the American tour. In fact, Mr Riddles had to take over the driving duties for much of the tour because of Bishop's illness.

NRM

Plate 36: Mr R.A. Riddles (left) receiving an American locomotive whistle from the Model Railroad Builders of America at Chicago on 2nd April 1939. The other train is the 'Capitol Limited' of the B&O Railroad.

R.A. Riddles collection

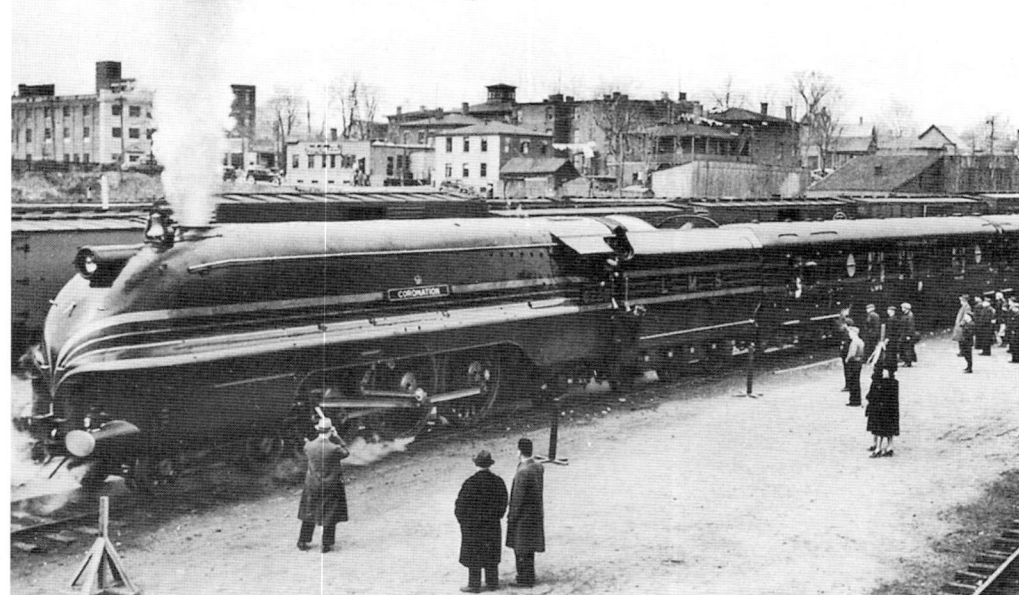

Plate 37: 'Coronation Scot' tour train at Hartford, USA in 1939. But for the war, complete new sets of coaches of this style would have gone into service on the LMS in 1940.

Photomatic

Plate 38: The complete train photographed at speed at East Chatham, New York State, USA in 1939.

Photomatic

Plate 39, 40: While the real *Duchess of Hamilton* was in the USA, the erstwhile *Coronation*, still painted blue, ran in Britain as No. 6229 and the adjoining pictures bear witness to this. *Above*, the engine is seen near Crewe on a routine express while *to the right*, the pseudo No. 6229 actually heads the 'Coronation Scot' express in August 1939. That it is clearly the original No. 6220 can be established by the flattened casing above the cylinders — a feature only found on the first five engines which were the only ones fitted with cylinder by-pass valves.

Real Photographs, Photomatic

Plate 41: In due course, the real No. 6229 returned home to resume its old identity and the picture alongside shows the engine in drab wartime black and its tender partially de-streamlined. Note the differences in detail of the front lower streamline casing, compared with 40 (above).

Real Photographs

THE BLACK STREAMLINERS

In 1943, the LMS bowed to the inevitable wartime circumstances and painted its last four streamliners (Nos. 6245-8) in plain black and soon repainted the others to match. Frankly they looked awful and it was no surprise when the company finally abandoned streamlining with the last batches to be built (No. 6249 onwards). In the meantime, the streamliners looked more and more woebegone, made even worse when the LMS started to remove bits of the casings for ease of access. At first it was the tenders which suffered. The extended sidesheets were cut back at the rear, followed (later) by the removal of the lower tender side valances. Finally, in 1945, the company decided to remove the engine casings altogether thus precipitating sighs of relief all round from the maintenance crews! No. 6235 was the first but by nationalisation, only two streamliners were left — No. 6229 (now returned from the USA) and No. 6243 which actually received its BR number before de-streamlining.

Plate 42-4: Left (above), No. 6243 *City of Lancaster*, with its tender partly de-streamlined heads a typical massive wartime train (c.16 or 17 bogies) while *below*, an even filthier No. 6244 *King George VI* is seen at Euston in 1946, about to leave with the West Coast postal. Lastly, a slightly cleaner, but still pretty uninspiring *City of London* forms the backdrop for a group of Crewe pensioners at the works centenary in 1943.

Real Photographs, BR (LMR), NRM

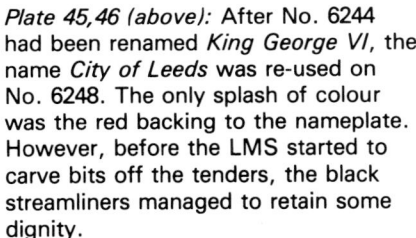

Plate 45,46 (above): After No. 6244 had been renamed *King George VI*, the name *City of Leeds* was re-used on No. 6248. The only splash of colour was the red backing to the nameplate. However, before the LMS started to carve bits off the tenders, the black streamliners managed to retain some dignity.

NRM, Real Photographs

Plate 47: Life was clearly a problem during the war. In the adjoining view, No. 6224 *Princess Alexandra*, painted black, has been paired with a non-streamlined tender from the No. 6230-4 series of engines. The tender, however, still retains its red livery as far as can be ascertained.

Real Photographs

Plate 48: The only streamliner to receive BR number was No. 46243 *City of Lancaster* and it is seen here in 1948 on a down Glasgow train.

Real Photographs

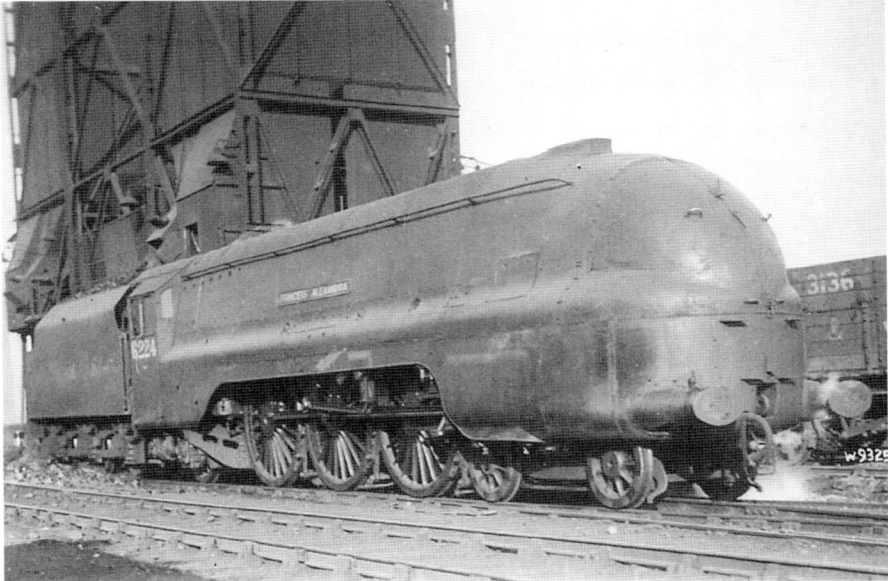

THE ORIGINAL NON-STREAMLINERS

William Stanier was utterly indifferent to the streamlining of his great engines but agreed to the move in order to satisfy the publicity department which, presumably, was becoming a little restless at all the attention showered upon the LNER streamliners. That the LMS was actually running far more trains at a start to stop average of 60mph or more than virtually the whole of the rest of the railway system added together was of no consequence in this particular context. However, having given the publicity men their ten streamliners, Stanier then built five more without casings 'for comparative purposes' as it was euphemistically stated at the time. One suspects that Stanier had his tongue in his cheek, for in these engines (Nos. 6230-4) one could see for the first time the truly noble proportions of the design. Significantly perhaps, in retrospect, all five (like the first red streamliners) were named after Duchesses — Buccleuch, Atholl, Montrose, Sutherland and Abercorn. These five captured the imagination in a manner that the streamliners never did and it was one of them No. 6234 which first revealed the real power potential of the design. Many writers, including such well known authors as Hamilton Ellis and Dr. Tuplin have averred that in these engines was epitomised one of the visual high points of British steam design. Need one say more, they were magnificent. . .

Plate 49: A classic Eric Treacy shot of No. 6232 *Duchess of Montrose* heading south at Penrith.

Plate 50,51: These posed ¾ front views show the first two non-streamlined Duchesses in original guise. *Duchess of Buccleuch* is seen at Shrewsbury when new and *Duchess of Atholl* at Perth in August 1939.
Real Photographs, Gavin Wilson

Plate 52 (below): At first sight it seems as if a non-streamlined Duchess is hauling the 'Coronation Scot' — until you count the number of coaches. It is in fact the Saturday 'Mid-Day Scot' (reporting No. W97) running as at least two separate trains and using the 'Coronation Scot' set as part of the formation of this, the first train — the locomotive is *Duchess of Atholl.*
N.E. Stead

Plate 53,54: Even when quite new, the Duchesses were not always as clean as desirable and the view of No. 6232 *Duchess of Montrose* at Edge Hill (*above*) depicts a distinctly grubby engine. However, *Duchess of Atholl* had clearly been well prepared before heading south with a Glasgow — London train, seen *at the left* passing Oxenholme, the junction with the Windermere line.

Eric Treacy, Real Photographs

Plate 55-7: The two views of No. 6233 *Duchess of Sutherland (above)* show the engine before and after fitting with double chimney. The engine was still in red livery at the time and remarkably clean, unlike *Duchess of Atholl* (*right*) which is seen here with double chimney and painted black. It is paired with a cut-back streamlined tender — probably from *Princess Alexandra* — see *Plate 47*.

Gavin Wilson collection, BR (LMR), Real Photographs

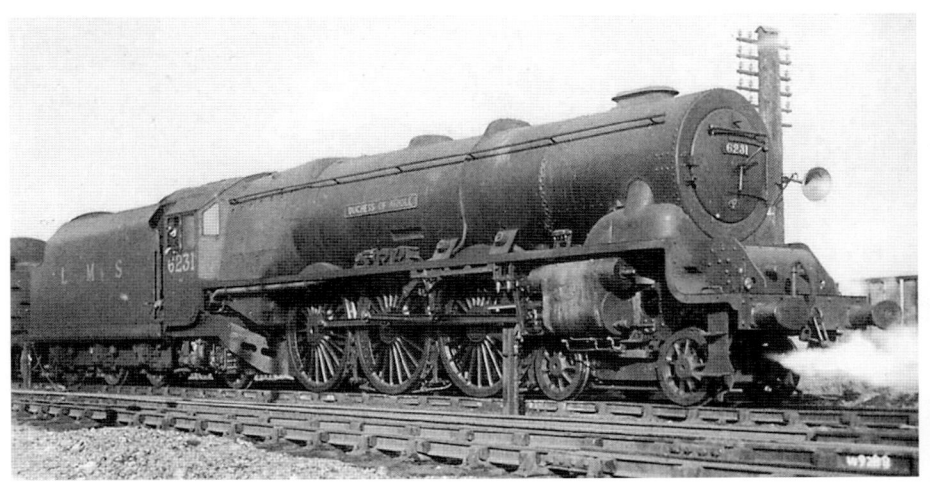

THE LAST NINE LOCOMOTIVES

Trials with No. 6234 in 1939 had shown the advantage of a double chimney which became standard from No. 6235 onwards and was applied retrospectively to all the earlier examples as well. Thus the final nine non-streamlined locomotives were all given double chimneys from new. The initial four (Nos. 6249-52) were actually ordered as streamliners and for a while ran with streamlined tenders built before the engines themselves. All were painted unlined wartime black but when No. 6253 appeared, the LMS had adopted its post-war livery — glossy black with maroon and straw lining. There were other changes too. From 1945, smoke deflectors had been fitted to the locomotives (other than to the residual streamliners). This was considered advantageous with the double chimney arrangement and became a standard feature from No. 6253 onwards, as did the interruption in the front footplating. This feature known as the "utility front", had first appeared on the de-streamlined examples and enabled the outside piston valves to be withdrawn more easily for examination. This modification was not applied retrospectively to Nos. 6230-4, 49-52 which retained the continuous front footplate to the end. Oddly enough, the de-streamlined No. 46242 *City of Glasgow* was given this continuous arrangement when it was extensively repaired (virtually rebuilt, in fact) after the disastrous Harrow double collision (*see also plates 147, 148*).

Finally, No. 6256 and No. 46257 (the last two built) were given modified rear trucks, roller bearings and other mechanical modifications. This involved fitting an attenuated cab side sheet which, in the opinion of the author, rather spoiled their appearance.

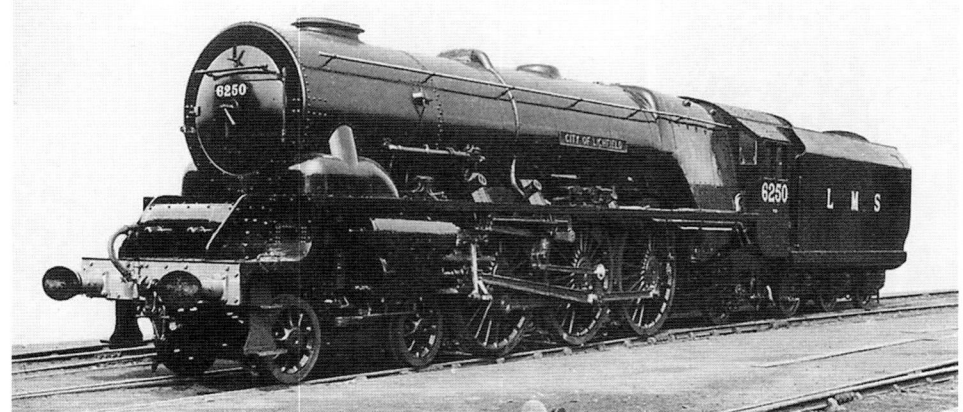

Plate 58,59: City of Lichfield is shown in two forms in these pictures. *At the left* is a view of the engine as built with streamlined tender and in wartime colours while *below*, the same engine now carries the 1946 LMS livery, is fitted with smoke deflectors and has a fully de-streamlined tender.
BR (LMR), Gavin Wilson

Plate 60,61: Superficially, there is little difference in these two views of *City of Stoke-on-Trent* and *Sir William A. Stanier, FRS* at Carlisle early in 1949, but the slightly different version of the 'utility' front and the shallower cab side sheets of No. 46256 are the main visible differences.

Gavin Wilson

THE DE-STREAMLINED LOCOMOTIVES

When de-streamlined, the Duchesses (now all with double chimneys), as well as receiving smoke deflectors, retained their cut-down smokebox tops for a few years. This visual obscenity did little for the looks of the engines — in fact, it is surprising to realise how such a minor difference could so seriously affect the looks of a locomotive. Some contemporary observers referred to them as 'semi-streamlined' in this form, nastily abbreviated to 'semis' by some platform end enthusiasts. How on earth a smokebox top which looked as though it had been attacked with an over size warding file could be in any way streamlined was a cause of some bafflement to the rest of us — but we digress!

The first engine to lose its casings was No. 6235 *City of Birmingham* shown *at the left* in works grey livery with two alternative chimney casings, the stovepipe version of which being, thankfully, quickly suppressed. These views show the earlier form of the LMS 1946 livery with no boiler lining and footplate lining confined to the lower edge. *City of London, below*, displays the later 1946 standard scheme on 16th February 1948 as it heads the centenary 10.00 a.m. express from London to Scotland.

NRM

Plate 65,66: The two pictures on this page show the de-streamlined engines *City of Leeds* and *City of Liverpool* on the 'Mid-Day Scot' duty. *City of Leeds* is seen about to leave Glasgow Central on 17th February 1948 while *City of Liverpool* makes a spirited southbound climb to Shap a year or so later with the same train — some of the coaches now in BR livery.

Gavin Wilson, Eric Treacy

THE FINAL PHASE

In due course, smokebox replacement caused the ex-streamliners to assume a 'proper' locomotive shape and in their final years from the mid-1950s onwards, all members of the class were at last recognisably from the same stable. There were still detail differences (footplate fronts, tender detail variations and the like) and the full gamut of variable colour schemes which characterised these engines — of which more in proper time — had still to run its full course. But in their final years, all the Duchesses conformed, more or less, to the classic lines first revealed in 1938. Some folk did not like the smoke deflectors (and there is some evidence that they were not strictly necessary) but, and again speaking personally, the writer's view is that the smoke deflectors gave a better visual balance to the front end of the locomotive by concealing the somewhat bulbous steam pipes which, although very efficient at getting steam in and out of the cylinders, were not very attractive in appearance.

Plate 67,68: Duchess of Montrose in BR green and *City of Liverpool* in BR red livery illustrate the visual similarity between the original non-streamliners and the de-streamlined engines in their final years. Note the downward extension of the smoke deflector plates on No. 46232 — a feature characteristic only of those engines carrying the full footplate front.

Eric Treacy

Plate 69,70: These two superb shots by Bill Anderson have probably been published before — if not, they should have been! Not only do they truly symbolise the power of the design, but they also clearly show the final BR configuration of the post-war non-streamlined engines. *City of Hereford* (*above*) heads the down 'Royal Scot' on Beattock bank in early BR days while *City of Salford* (*below*), the only BR built example of the class and the last to be built, is in charge of the down 'Caledonian' a few years later.

Plate 71,72 (left and above): Blue streamlined No. 6224 *Princess Alexandra* and single chimney red non-streamlined No. 6232 *Duchess of Montrose* at Shrewsbury in 1937 and 1938 respectively. *Steam and Sail*

LOCOMOTIVE CHAMELEONS

It is at least arguable that no class of British steam locomotive exhibited such a wide variation of exterior styles and colour schemes during such a brief period (basically only 20 years) as did the Duchesses. From the appearance of the streamlined No. 6220 in 1937 to the final BR red livery assumed by some of them from 1957 onwards, no fewer than 12 basic liveries and five different external forms were exhibited by the class. Astonishingly, only one livery (BR green) was actually carried by all members and at no time did all the engines look exactly alike in all respects. On the next pages an attempt is made to summarise the story by providing pictures of all the principal guises in which they appeared. For obvious reasons, this seems the appropriate page to introduce colour pictures.

Plate 73 (above): No. 6231 *Duchess of Atholl* is seen in this view at Crewe, carrying the early BR standard blue livery in 1952. *W. Hubert Foster*

Plate 74 (left): This rare colour shot shows de-streamlined *City of Edinburgh* carrying the experimental BR blue livery with LNWR lining in 1948. *Steam and Sail*

Plate 75 (below): City of Lichfield in BR standard green with the later tender emblem. Although all engines carried this colour scheme during the mid 1950s, it is not certain whether all 38 were given it with both versions of the emblem. *Steam and Sail*

Plate 76-9: **Variations of the BR red livery.** Only 16 examples were painted red by BR but to the writer it was the only proper colour for a Duchess — just as dark green was the only right shade for a GWR 'King'. *Above left* is shown a cab-end view of *City of Liverpool* wearing the short lived BR style of lining while *above right*, the standard version is shown in close-up on *Duchess of Hamilton* while under restoration at Swindon. *To the right and below* are seen *Sir William A. Stanier, FRS* at Carlisle and, perhaps most symbolically of all *City of Nottingham* alongside *Hereford Castle* at Swindon in 1964. One hopes that Churchward would have approved!

Gavin Wilson, David Jenkinson, John Edgington

Plate 80,81: **Red and black streamliners,** *King George VI* (*above*), when renamed from *City of Leeds* and (perhaps?) *City of Birmingham* below — note the paper numbers stuck to the cabside — the LMS up to its old tricks again!

NRM

Plate 82,83: The 1946 livery — both engines in photographic grey. *City of Birmingham* — this time the real one (!) shows the earlier form of lining and lettering while *City of St Albans* shows the later lettering and modified footplate and cylinder lining.

NRM

Plate 84,85: City of Glasgow (above) displays the full 1946 livery with lined out boiler bands while *City of Liverpool (right)* displays one of the many hybrid forms visible in the early BR period — this time the second version of the LMS 1946 livery but with BR totem on the tender — a rare combination.
Real Photographs, Gavin Wilson collection

Plate 86 (below): Duchess of Buccleuch in experimental BR blue with LNWR lining.
BR LMR

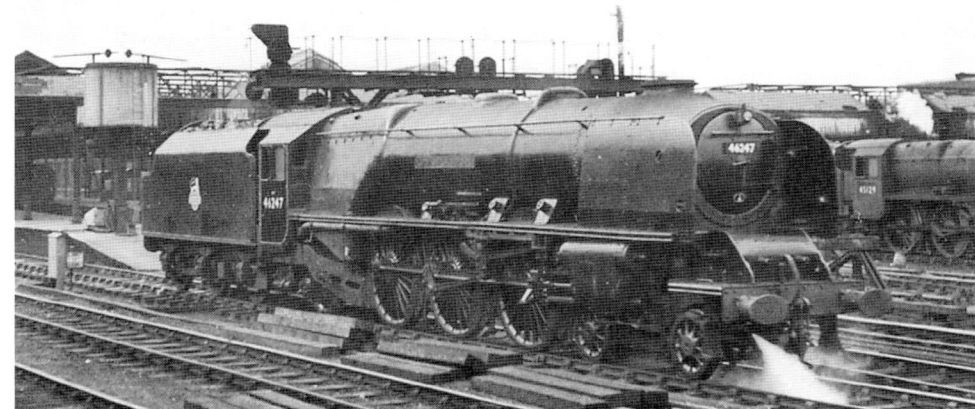

Plate 87,88: City of Leicester in LNWR style experimental black *(right)* and *City of Lichfield* in BR standard blue *(below)*. In the view of the writer this latter colour scheme sat rather well on the Stanier Pacifics — far better than the green which succeeded it in 1951-2.
Gavin Wilson, BR LMR

DUCHESSES IN DETAIL

The Duchesses have long been a popular prototype for modelmakers, including the author. It therefore seemed no bad idea to include a few pages of pictures in which these aspects receive attention.

Fronts — I: These four views show the front ends of the Duchesses in the pre-war period. *Above* can be seen the streamlined arrangement, both open and closed — No. 6220 on the left, No. 6225 on the right. Note the extra steps on No. 6225. These were fitted retrospectively to Nos. 6220-4 almost immediately after the engines were completed — see *Plates 16-19*. *Below* are contrasted the streamlined and non-streamlined series from the same angle of view.

NRM

Fronts — II: Apart from 93 (*above, left*) which shows No. 6230 when built, the other three pictures show Duchess front ends in post-war configuration. The lack of riveting at the base of the smokebox of No. 46245 in original de-streamlined state (*above, right*) is worthy of note.

NRM, Real Photographs, Gavin Wilson, BR LMR

From the cab end: *Above* is the cab view of No. 6230 when new, while *to the left* are views of three members of the class from almost the same angle. Note the gradual increase in complexity from the early view of de-streamlined No. 46241 (*top left*) to the immediate pre-scrapping view of No. 46256 (*bottom left*). *Below right* is a rear view of No. 6224 showing, amongst other features, the first style of post-war insignia used by the LMS.

Real Photographs, NRM, Alex McNair, J.R. Carter

Plate 102, 103 (above): Boiler top details on No. 46233 (*left*) and 46227 (*right*) at Carlisle.

Eric Treacy

Plate 104, 105 (below): These two views from the footplate of No. 46240 *City of Coventry* climbing Shap and No. 46257 *City of Salford* near Manchester give a very clear impression of the huge boilers of the Duchesses.

Eric Treacy, J.R. Carter

Plate 106, 107 (above): Wheel and nameplate detail of streamlined *City of Birmingham* and de-streamlined *City of Bradford* in 1944 and 1948 respectively.

NRM, BR LMR

Plate 108-11 (below): The footplate top fittings above the driving wheels of the Duchesses were sometimes collectively referred to as the 'Plumber's Yard'! Three of the accompanying views illustrate the items provided. Close scrutiny will reveal that they are all different in detail. The upper view of *Coronation* shows how much more difficult it was to get at the fittings in the streamlined state.

NRM

Tenders Plates 112-14 (above and right) show three views of the original streamline pattern tender fitted to the Duchesses — the steam operated coal-pusher is visible through the tender front door (top right). On the left, below, is the cut back streamline tender of No. 6241; while below right is the post-war non-streamlined pattern tender of No. 46254.

NRM, BR LMR, J.R. Carter

DUCHESSES ON TEST

The high speed nature of the 'Coronation Scot' service perhaps, at first, tended to obscure the fact that the Duchesses were designed as much for heavy haulage as for lightweight fliers. However, in 1939, *Duchess of Abercorn* was put to the test and indicated over 3300hp at the cylinders — a figure which established the 'most powerful' claim for the class. Similar figures were established in 1955-6 with *Duchess of Gloucester* when, on one trial run she topped Ais Gill summit at 30mph with the equivalent of *900 tons* on the drawbar. These exploits are well recorded in print, as are those of *City of Bradford* in the 1948 locomotive exchanges. The latter event seems to have disappointed Duchess admirers yet, if truth be told, none of the express types performed throughout as they could at their best. But the Duchess was only marginally in second place on fuel consumption and was the only locomotive tested in its class which fulfilled all its assignments without either change of locomotive or mechanical failure — Stanier's insistence on easy maintenance obviously paying off here. The pictures on these two pages provide a photographic reminder of these events.

Plate 117, 118 (above and left): It is a moot point whether the *Duchess of Gloucester* or the *Duchess of Abercorn* trials represented the peak of test performance of these engines. These two views show *Duchess of Gloucester* festooned with test equipment at the Rugby test plant in 1955 on the occasion when the highest continuous steaming rate (in excess of 40,000 lbs/hr) for any British express locomotive was achieved.

Alex McNair, John Edgington

Plate 119 (below): In 1939, *Duchess of Abercorn* was tested on the road — with both single and double chimney — with loads above 600 tons between Euston and Glasgow. This official view was taken almost immediately after the fitting of the double chimney.

NRM

Plate 120, 121 (above, left and right): City of Bradford *on East Coast territory. At the left she is seen receiving preliminary attention at Kings Cross on 6th May 1948 while on the right, the engine is seen at New Barnet on 29th April.*
BR LMR, Real Photographs

Plate 122 (right): Sonning Cutting and the 1.30 p.m. Paddington-Plymouth are the location and train in this view of No. 46236 on the Western Region.
M.W. Earley

Plate 123 (below): Lack of water troughs on the Southern Region necessitated the fitting of a tender with higher water capacity when City of Bradford *was rostered to the 10.50 a.m. Waterloo-Exeter train in July 1948.*
BR LMR

Plate 124 (above): Red painted but somewhat grimy, No. 46251 *City of Nottingham* prepares for duty at Camden in March 1962.

Steam and Sail

Plate 125 (above): Carlisle based *City of Bradford* takes a well earned rest under the arches of old Euston station after a 300 mile run.

John Edgington

Plate 126 (left): City of Carlisle slips down the south side of Beattock bank with the 13.45 Glasgow-Liverpool on 29th July 1961.

Don Rowland

Plate 127 (below): The driver of *City of Chester* completes his examination of the engine prior to departure from Carlisle with the up 'Mid-Day Scot.'

Gavin Wilson

Plate 128, 129 (above): City of Manchester waits the 'off' at Euston (left) while City of London, surrounded by evidence of her electrified successors-to-be is relegated to freight duties at Crewe.

John Edgington

Plate 130, 131 (above): Sir William A. Stanier, FRS was the last of the class to run in traffic. Often called upon for excursion work in 1964, the engine is seen in close-up (above), while to the right is a lovely shot of her climbing Shap for the very last time with an RCTS excursion in September 1964. This hardly looks like a locomotive ready for the scrap heap the following week, does it?

Gavin Wilson, Steam and Sail

DUCHESSES AT WORK

So far, attention has been mainly concentrated on the chronological story of the engines. But no matter what the external style or colour scheme, the purpose of a locomotive is to work trains. Let us now, therefore, take a closer look at the engines (in all their guises) at work in their prime . . . and by way of a curtain raiser — two pages of colour pictures.

Plate 132, 133 (right): City of Hereford was another excursionist in 1964. On 12th July she headed a southbound excursion over the Settle and Carlisle and is seen being prepared at Kingmoor (above) and passing Dent Head (below). Those were the days when one could still photograph steam without half the nation present — I had Dent Head to myself!

Gavin Wilson, David Jenkinson

Plate 134: Euston was a fascinating if impossible terminus in the days before the official vandals knocked it all down and at least the Duchesses were spared the indignity of having to witness *that* happening! Here, *Duchess of Atholl* gets under way with the down 'Royal Scot' in the 1950s.

BR LMR

Plate 135: Almost twenty years earlier, the most frequently seen Duchesses were the blue streamliners. The exhaust emanating from the single chimney of No. 6223 *Princess Alice* suggests that there was going to be no messing about with the down 'Coronation Scot' on this occasion.

Photomatic

Plate 136 (above): During the late 1950s, the West Coast main line saw the introduction of a new flier — the 'Caledonian'. It was not a heavyweight train but its timing needed Duchesses at their best. In this view, No. 46226 *Duchess of Norfolk* pulls out with the down train c.1960.

Eric Treacy

Plate 137, 138 (right): The old arrival side at Euston with its sweeping curved platforms and canopies could always be relied upon to produce a few Duchesses at almost any time of day or night. On 25th March 1963, the spring sunshine makes a lovely pattern of light and shade on *Duchess of Rutland*, just arrived with the 08.00 from Blackpool.

John Edgington

Plate 139 (above): When the railway first came to Euston, Primrose Hill probably lived up to its name but when the Duchesses held sway, it was merely the location of a fascinating tunnel complex. Here, No. 46225 *Duchess of Gloucester* heads north with a heavy express in 1951.

Real Photographs

Plate 140 (below): The Duchesses were worthy successors to a long line of famous LNWR performers and the picture of No. 6222 *Queen Mary* passing under the old LNWR gantry at Willesden in April 1939 is nicely symbolic. But which signal applied to the train?

Gavin Wilson

Plate 141, 142: Camden Bank was a favourite haunt of railway photographers. To the right, *City of Liverpool* backs down towards Euston to take charge of the 'Ulster Express' while *below*, *City of Leeds* heads north with a down express on the adjoining main line.

Eric Treacy

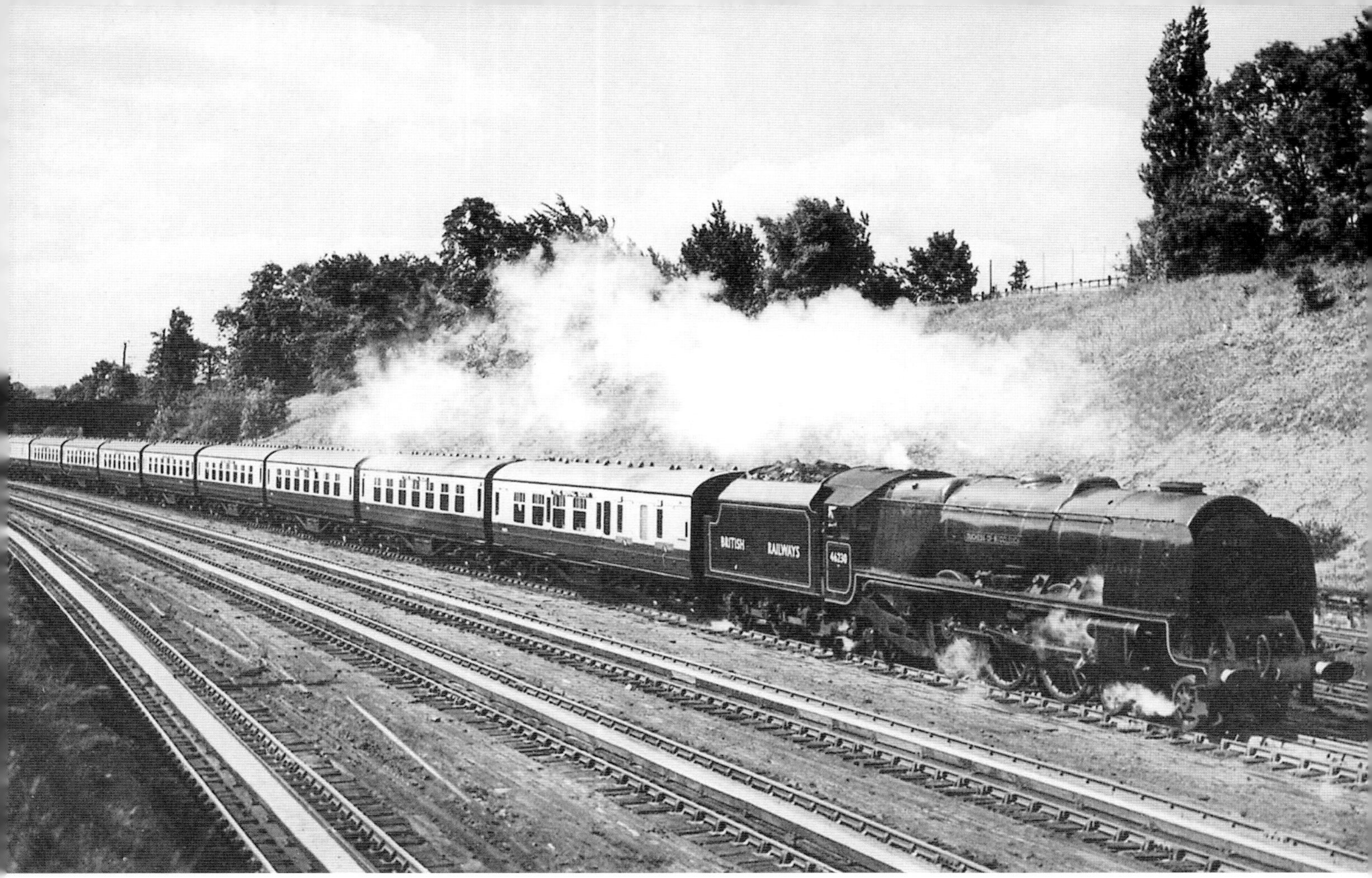
BRITISH RAILWAYS

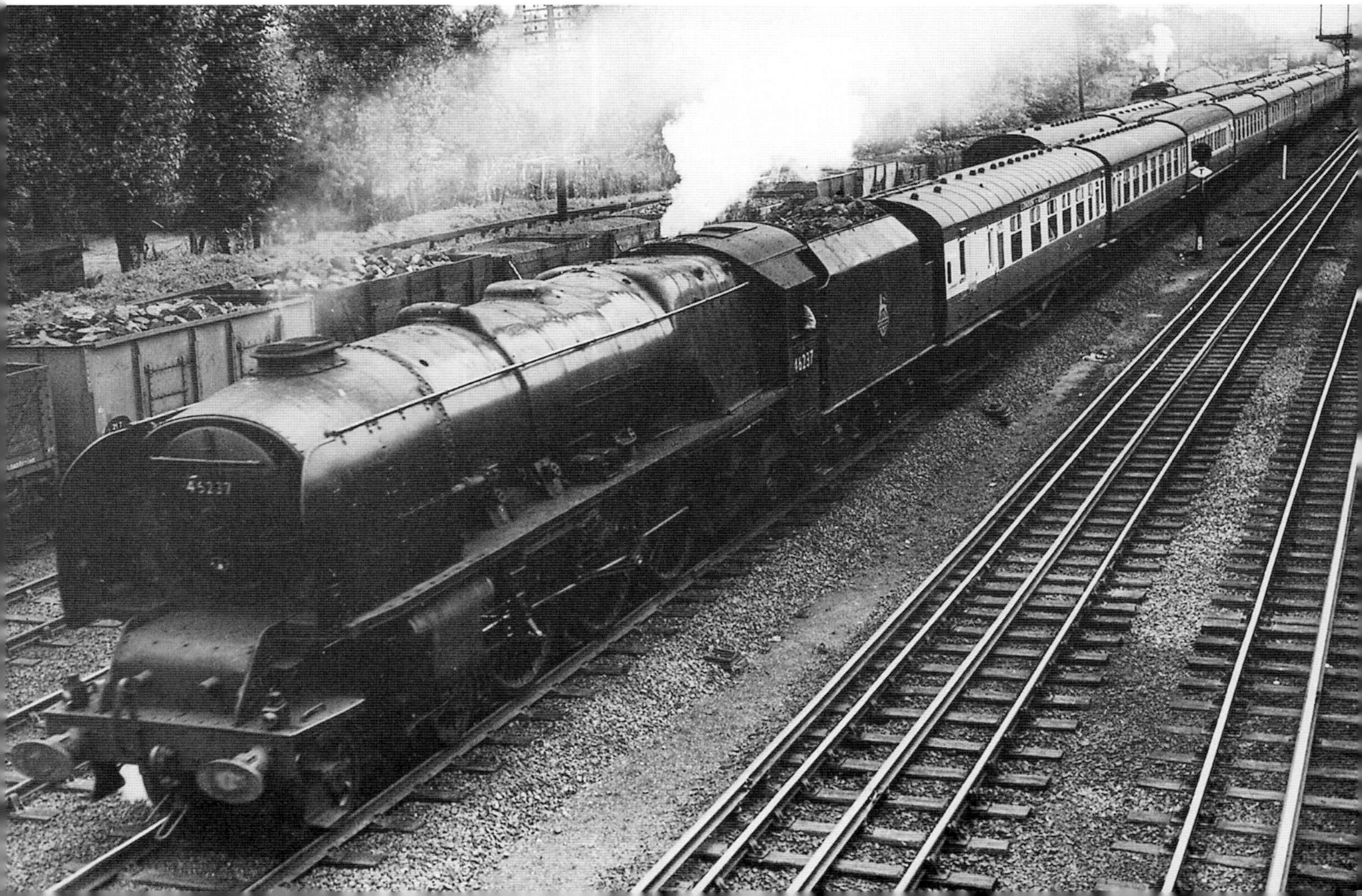
45237

Plate 143, 144 (opposite): These early BR views show the changing colour scene in the London suburbs. *Above*, No. 46230 *Duchess of Buccleuch* is painted in the 1948 experimental dark blue and is hauling 'plum and spilt milk' coaches on the down 'Royal Scot' at Bushey troughs. *Below*, green painted *City of Bristol* heads the red and cream Lakes Express past Hatch End.

BR LMR

Plate 145, 146 (this page): Later scenes along the London approaches in June 1960. *Above*, No. 46228 *Duchess of Rutland* heads south past Kenton with a train from Carlisle while *below*, No. 46225 *Duchess of Gloucester* is at Denbigh Hall, near Bletchley with the up 'Caledonian'. Engines and trains are both now painted maroon.

Derek Cross, BR LMR

Plate 147, 148: City of Glasgow was the only survivor of the three engines involved in the frightful Harrow collision of 1952 and as a consequence became unique in another far less macabre way. Before the accident the locomotive was in the conventional de-streamlined form as seen in 1947 at Castlethorpe *(left)*. On return to traffic, after virtually total reconstruction, the engine emerged as the tenth member of the class to sport the full front footplate, shown clearly in the view *below* of the Ulster Express at Watford in 1959. She was thus the only member of the class to carry all three variations of front end (streamlined, utility and non-utility).
Real Photographs, BR LMR

Plate 149: Not much coal left in the tender of *King George VI* as it passes Apsley with the up 'Caledonian' c.1957. This picture gives a very clear indication of the extent to which the Duchesses made maximum use of the structure gauge, compared with the noticeably lower height of the coaches.
BR LMR

Plate 150, 151: These two pictures, both taken at Roade on 25th May 1960, afford another opportunity to compare the front end differences in detail between members of the class. The upper view shows *Sir William A. Stanier, FRS* in charge of the 'Royal Scot' while the lower picture depicts *City of Sheffield* heading the 'Caledonian'. Both trains are heading south.

BR LMR

Plate 152: A much earlier view of the up 'Royal Scot' is represented by this picture of the north end of Stafford station c.1950. The engine, in BR blue livery, is No. 46221 *Queen Elizabeth* and the train is at least 15 coaches long.

Real Photographs

Plate 153-5: Contrasts around Crewe. *Above, left*, an immaculate red streamliner *Duchess of Devonshire* picks up the Penzance-Glasgow through coach in 1938 while to the *right*, some ten years later, a somewhat less pristine *Queen Mary* awaits its next duty at the same location. The view *below* shows *City of Coventry* heading south from Crewe c.1950 with an unidentified express. Note the handkerchief tied on the fireman's head!

Real Photographs, Don Rowland, Eric Treacy

Plate 156: A nicely symbolic picture, this one. It shows *City of St Albans* heading west through Eccles on the Liverpool-Manchester main line — the epitome of British steam on the world's oldest Inter-City route.

J.R. Carter

Plate 157: In this view, Carlisle based *City of Bristol* is homeward bound with a northbound freight at Wigan in the early 1960s. Freight working was to become increasingly the lot of the Duchesses in their latter years but at least on this occasion, the load seems worthy of the engine.

J.R. Carter

Plate 158: In general, the writer is not over enthusiastic about atmospheric pictures which show little or no detail of the train but this late evening view of *Duchess of Montrose* heading into the dusk with a Birmingham-Glasgow express at Golbourne No.1 Box has a nice evocative quality and serves as a reminder that much of the best Duchess work was performed during the hours when neither photographers nor footplate observers were present.

BR LMR

Plate 159: Another evening shot, but this time in summer, shows *City of Bradford* with the down 'Caledonian' at Winwick Junction on 21st August 1957.

L. Elsey

Plate 160, 161: The climb out of Liverpool Lime Street up to Edge Hill was always a demanding job for a locomotive — even a Duchess. The late Bishop Treacy made this area very much his own, photographically and on this page are two characteristic views. *Above*, City of Stoke-on-Trent is still blowing off vigorously as she tops the bank with at least twelve bogies tied on the drawbar while *below*, City of Salford heads into the murky tunnels off the platform end at Lime Street with an up express.

Plate 162, 163: The North Wales coast saw the Duchesses less commonly than the old LNWR main line proper but they were nevertheless, quite frequent visitors. On this page, *Queen Mary (above)* and *Sir William A. Stanier, FRS (below)* were photographed at Llandudno Junction and the approaches to Chester respectively.

J.R. Carter

Plate 164-6: Bangor is at the opposite end of the North Wales main line to Chester and Bangor No. 2 box was situated close to that characteristic tunnel mouth of quasi-Egyptian styling. In the *top and bottom views* on this page, *Duchess of Gloucester* is seen both passing the box and heading into the tunnel with a lightweight Holyhead-Crewe train while the *centre view* is from the footplate of *City of Bristol* waiting to be released from the loop.

Gavin Wilson collection

Plate 167, 168: Although Crewe was always the locomotive changing point on the Euston-Carlisle main line, to the author, the town of Preston always seems to be the place at which the route really changes its nature. The northern section will have to wait for a few pages more but on this page, two very contrasting views of Preston will serve to bid temporary farewell to England. *Above, City of Stoke-on-Trent* makes a rousing start with a heavy southbound express in 1947 while *below*, the 'Royal Scot' passes northbound through the town in 1952 behind an unidentified Duchess.

Eric Treacy, BR LMR

Plate 169, 170: In Scotland, as in England, the Duchesses were essentially West Coast main line locomotives but, as was the case south of the border, they were also seen elsewhere. *Above*, No. 46224 *Princess Alexandra* leaves Mauchline on the GSWR main line with a Glasgow-Carlisle local in September 1963 only one month before her withdrawal while *below*, *Duchess of Montrose* seems almost dwarfed by the superb span of Ballochmyle viaduct in April 1962 with the Euston-Glasgow sleeper taking the GSWR route.

Derek Cross

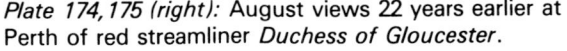

Plate 173 (above): This picture has been used before, but who could resist using it again? It shows *City of Edinburgh* in the evening sunshine in August 1961 leaving Perth (near Hilton Junction) with the 8.15 p.m. to Euston — this was what the Duchesses were all about.

W.J.V. Anderson

Plate 174, 175 (right): August views 22 years earlier at Perth of red streamliner *Duchess of Gloucester*.

Gavin Wilson

◁ *Plate 171, 172 (opposite):* Not just requiems for Duchesses but also for Glasgow termini. In the upper picture, *Queen Mary* rouses the echoes at St Enoch as she leaves with a Carlisle train in September 1962. Although the headlamps say 'express', the first four coaches are non-corridors. Below, the same engine is seen with the 'Royal Scot' leaving Buchanan Street during the period when the train used this terminal — Glasgow Central being in the throes of electrification.

Derek Cross, Gavin Wilson

DUCHESSES IN REPOSE

When all the emotion is stripped aside, locomotives were and are built to move traffic and on this score, the Duchesses were the most intensively used examples of any of the contemporary British express passenger types averaging some 7000 revenue miles a year more than their nearest competitors. However, even the most intensively used steam engines spent much of their time at rest and the next section of this survey shows them in more placid mood. Even at rest, however, the feeling of power was ever present.

Plate 176, 177: **Red Duchesses in London.** *To the left,* City of Coventry *turns at Camden while* below, *an immaculate* City of London *stands alongside the platform end at Euston. This engine was the first to be repainted red by the London Midland Region.*
Alex McNair, BR LMR

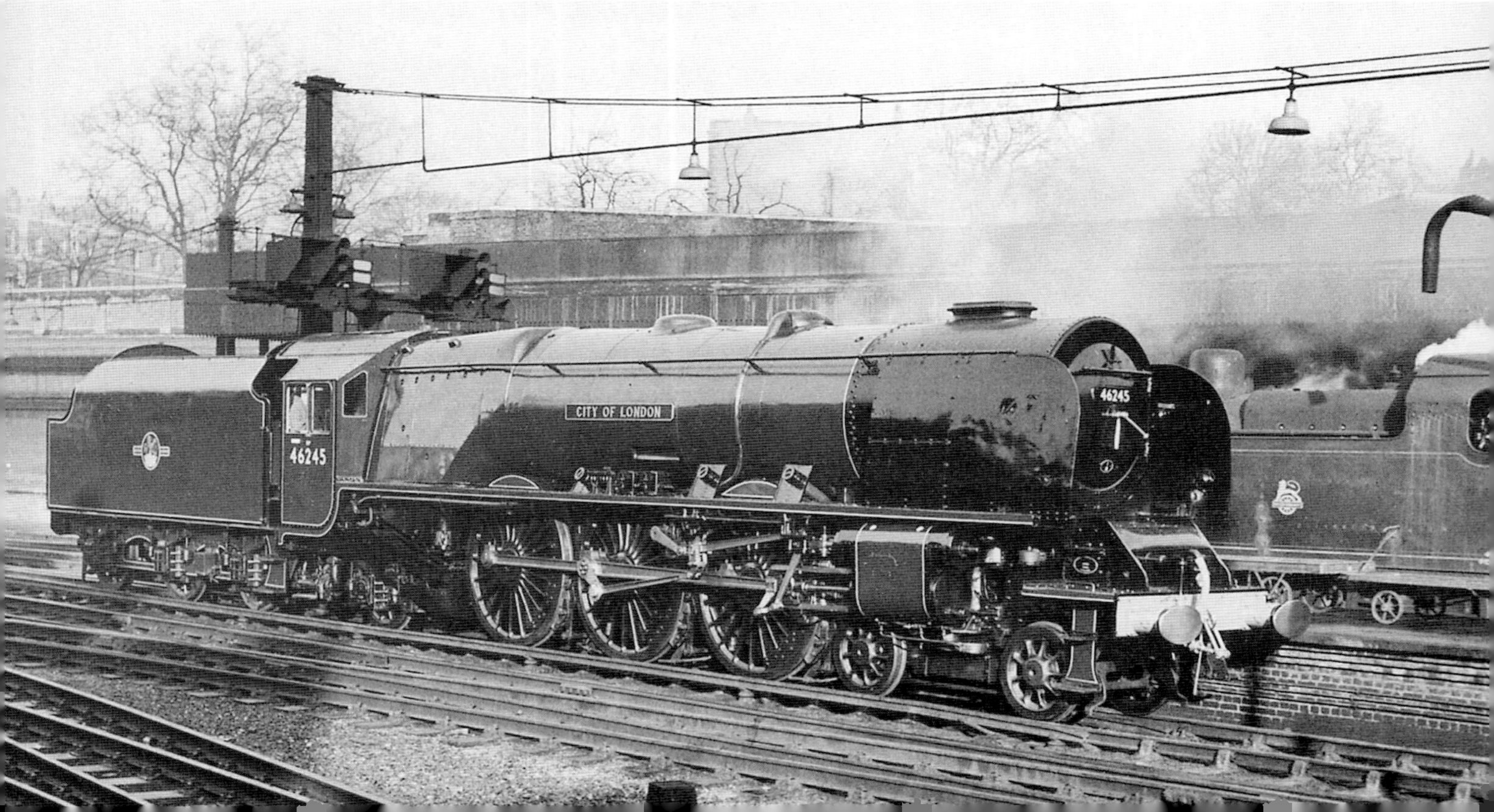

Plate 178, 179: **Interlude at Edge Hill** — *City of Stoke-on-Trent* and *Duchess of Hamilton* in LMS lined black and rather grubby BR red.
Eric Treacy, J.R. Carter

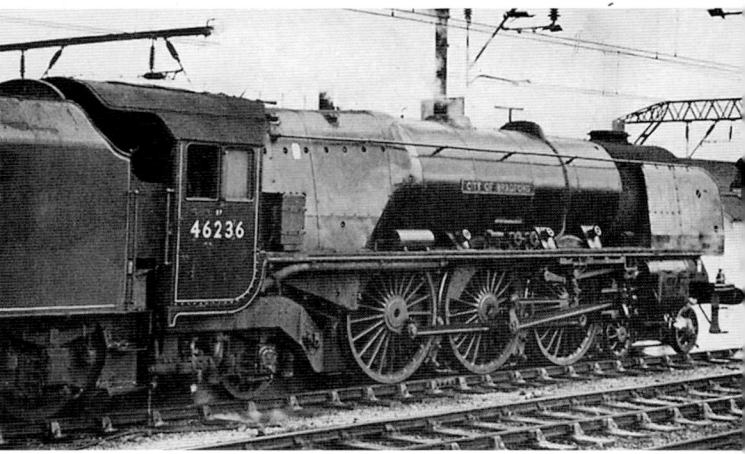

Plate 180, 181: City of Manchester and *City of Bradford* wait for the road at Crewe and Rugby respectively in the early 1960s with ample evidence of the forthcoming electrification all round. During 1964, the Duchesses were prohibited from working 'under the wire' south of Crewe and for a few months carried a yellow stripe on the cab sheet to denote this prohibition.

Gavin Wilson

Plate 182: City of Nottingham waits for the signal somewhere on the Western Region between Swindon and Banbury with an excursion train in 1964.
Gavin Wilson collection

Plate 183: The more compact and massive proportions of the Duchesses compared with their Princess Royal predecessors can be seen in this view of de-streamlined *City of Manchester* and an unidentified Princess at Crewe.
Gavin Wilson collection

Plate 184, 185: The Duchesses were built at Crewe, were shedded, inter alia, at Crewe and frequently changed duties at Crewe — so let us stay at Crewe for a while. *Above, City of Glasgow* is seen in front of *City of Nottingham* in early BR days while *below, City of Lichfield* waits under the smoke hoods at Crewe North MPD prior to a return trip to her home base at Carlisle.
Real Photographs, J.R. Carter

Plate 186-9: These four fine shed studies at Crewe North were all taken by Driver J.R. Carter. The locomotives concerned are *City of Chester*, *City of Coventry* (*opposite*) and *City of Stoke-on-Trent*, *Duchess of Rutland* (*this page*) — a truly majestic series of pictures by a dedicated railwayman.

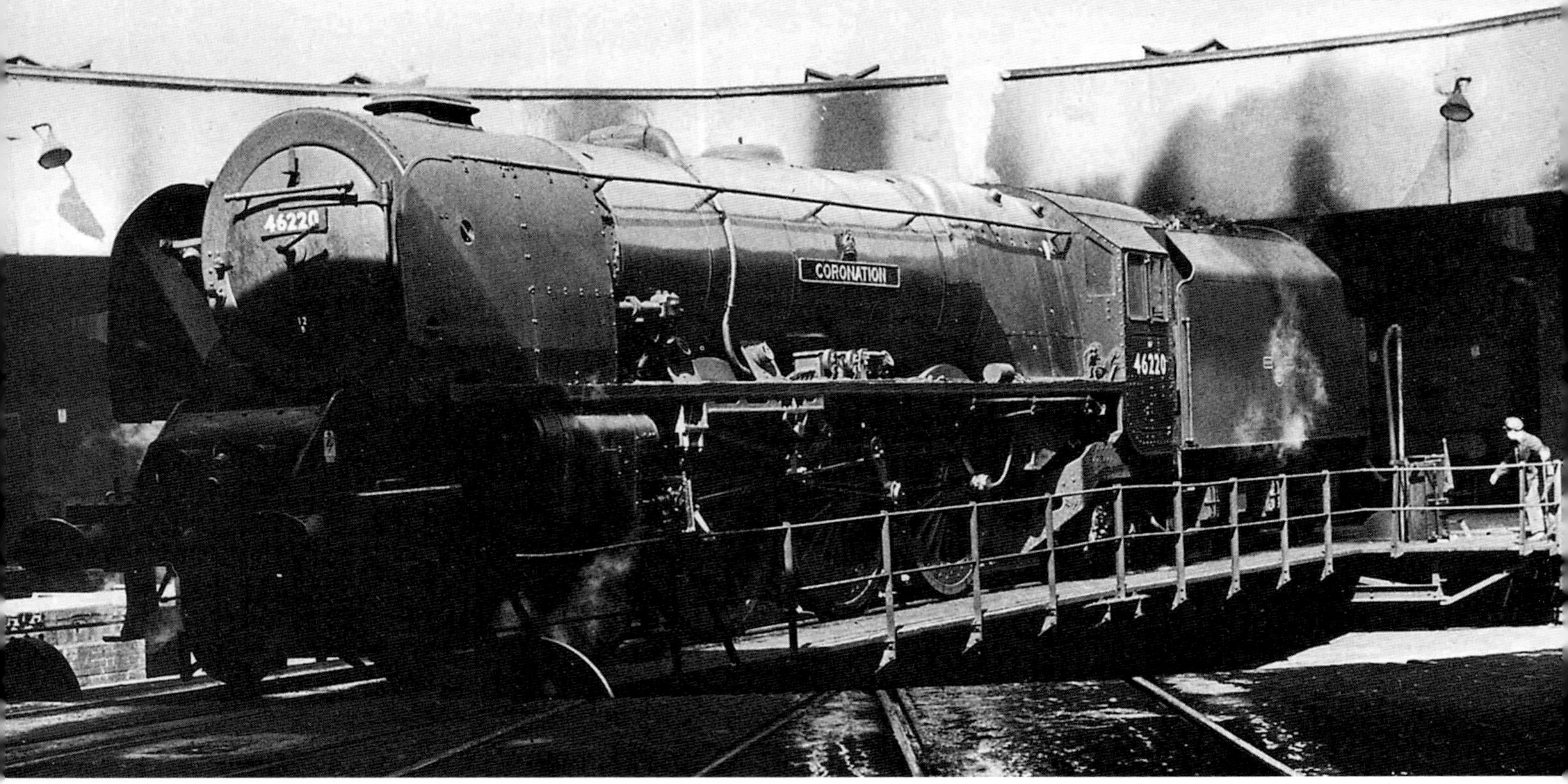
46220
CORONATION
46220

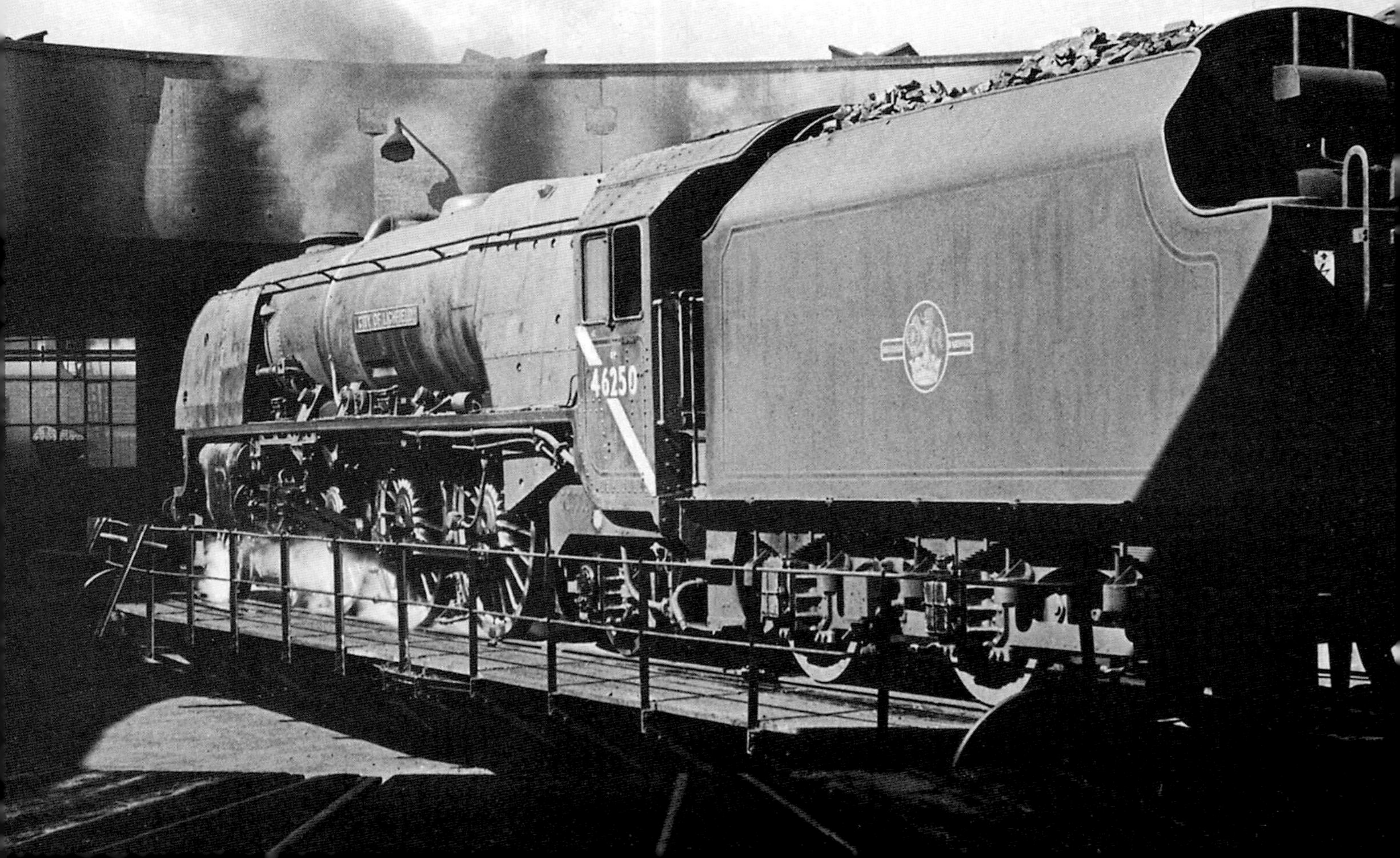
46250

Plate 190-3: Carlisle Upperby was another shed which saw a great deal of the Duchesses and it was also one of the many hunting grounds of the late Bishop Treacy. These four pictures show *Coronation* and *City of Lichfield* on the main roundhouse turntable (*opposite*) and leaving the shed for further duties (*this page*).

Plate 194, 195: Resting periods complete, two more Duchesses reverse down to their trains, *Duchess of Norfolk* at Carlisle and *Sir William A. Stanier, FRS* at Crewe. As with the previous few pages, these pictures were taken by Eric Treacy and J.R. Carter respectively.

DUCHESS COUNTRY — NORTH OF PRESTON

Many engines seemed particularly 'at home' in certain specific areas and for me, the real 'Duchess Country' was the main line over the hills between Preston and Glasgow. By way of an overture on this page are portrayed *Duchess of Abercorn* taking water from Brock troughs *(right)* and *City of Lichfield* heading north from Lancaster to do battle with the high hills *(below)*.
Photomatic, BR LMR

Plate 198-200 (left and below): North of Lancaster, the first real obstacle is Grayrigg bank and by Oxenholme (*left*), the gradient is showing. In these two views, red streamliners *Duchess of Devonshire* (*above*) and *Duchess of Gloucester* (*below*) are climbing and descending through Oxenholme respectively. The gradient is obvious by comparison with the Windermere branch to the left. The picture below at Hay Fell shows *Duchess of Gloucester* in much later days with the down 'Mid-day Scot' in August 1960.

Gavin Wilson collection, Derek Cross

Plate 201,202 (opposite): The Lune gorge provides a respite between Grayrigg and Shap, not to mention some beautiful scenery. In the *upper picture*, *City of Carlisle* heads north on a sunny July Saturday in 1963 with a Glasgow train while *below*, *City of Stoke-on-Trent* takes a good fill from Dillicar troughs prior to tackling Shap on 10th May 1956.

Derek Cross, David Hepburne-Scott

Plate 203, 204: Framed by the trees at Dillicar, *City of London* prepares to take water (*above*) in April 1962 with the morning Perth-Euston train while *City of Leeds* (*lower*) is obviously not bothering to replenish the tank as she passes over the troughs with a Glasgow train in the early 1950s.
Gavin Wilson, Eric Treacy

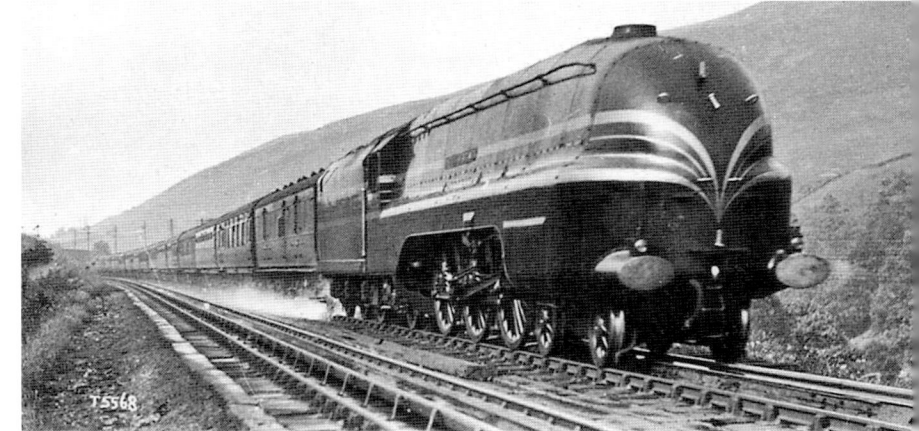

Plate 205-7: In this group of pictures, *City of Lichfield* heads south through Tebay in 1960 with the 'Caledonian', red streamliner *Duchess of Rutland* takes water with the up LMS 'Midday Scot' in 1939 and *Duchess of Buccleuch* does likewise with a southbound relief train in August 1960.

Derek Cross, Real Photographs

Plate 208-11: Tempting though it would be to fill page after page with views of Shap bank, these four will have to suffice at this stage — they hardly need captions. The locomotives concerned are *City of Bristol* and *King George VI* (*opposite*) and *City of Stoke-on-Trent* and *City of Birmingham* (*this page*). The latter view was taken right at the end of the Duchess story in August 1964 — at least 'fourteen on' and no banker. Not bad for a redundant engine!

W.J.V. Anderson, Eric Treacy, Derek Cross

Plate 212, 213: The southbound climb to Shap was less dramatic than the northbound ascent but in many ways a tougher proposition. Thrimby Woods was a popular location and *City of Coventry* makes a spectacular ascent with the up 'Royal Scot' in 1956 (*above*). *Below, City of Leicester* takes life a little easier at Thrimby Grange box with a Glasgow-Birmingham train in July 1957. Note the safety valves still just blowing off.

Eric Treacy, L. Elsey

Plate 214-16 (opposite): More scenes at Thrimby. In the *upper* picture, *City of Lancaster* still had steam to spare with almost 500 tons of train when photographed by Eric Treacy in the late 1950s while *below*, the down 'Royal Scot' and the up Perth-Euston were caught in quick succession by Derek Cross at this delightful pastoral location in August 1960. The locomotives are *Duchess of Buccleuch* and *Princess Alice* respectively.

Plate 219 (above): Duchesses were rarely seen double headed but when they were, it was forbidden to couple them as leading engine! *City of Sheffield* is seen breaking the rules at Carlisle in 1948.
Real Photographs

Plate 220 (right): This striking view shows *Duchess of Montrose* in experimental blue livery arriving at Carlisle with the down 'Royal Scot' c.1948-9. The coaches are also in experimental colours.
Real Photographs

◁ *Plate 217, 218 (opposite):* In the *upper view, Duchess of Hamilton* pulls out of the loop south of Penrith with a down parcels train in July 1963 while in the *lower view City of Glasgow* is getting a vigorous work-out with a Birmingham train on Easter Monday, 1961.
Derek Cross, S.C. Crook

Plate 221 (below): Three for the price of one at Carlisle on 16th June 1956. Left to right the engines are *Princess Alexandra* (arrived with the up 'Royal Scot'), *City of Lichfield* (about to take over the 'Royal Scot') and *City of Lancaster* on the Glasgow-Birmingham.
John Edgington

Plate 222-4: It was always worth while lingering at Carlisle in Duchess days. On this page, the two *upper views* show *City of Stoke-on-Trent* leaving Citadel station in April 1948 with a southbound train while the *lower view* shows *City of Carlisle* leaving her namesake city in the early 1960s.

Gavin Wilson, Eric Treacy

Plate 225 (above): For a while in the 1950s, the last two Duchesses to be built carried electric headlamps — hence the additional cables along the boiler side of No. 46256 *Sir William A. Stanier, FRS*, seen leaving Carlisle with a southbound train. It will, however, be noted that the old fashioned oil lamps are also present!

Eric Treacy

Plate 226 (overleaf): This seemed to me such a beautiful picture that it deserved a double page all to itself. It shows *Duchess of Atholl* about to take over the down 'Royal Scot' from *King George VI* at the north end of Carlisle station. Needless to say, it was taken by Bishop Eric Treacy.

Plate 227-30: **Northbound from Carlisle** The journey north resumes again with yet another clutch of Eric Treacy pictures. On this page, the *upper view* shows Polmadie based *Queen Mary* just getting under way with a northbound express — and being chased along by a group of schoolboys — while the *lower view* shows *Duchess of Norfolk* backing down to take over a northbound train not yet arrived. *Opposite (above),* blue painted *Coronation* sets out with a Birmingham-Glasgow train c.1950 while in the *lower view, Princess Alice* passes Carlisle No.4 box some ten years or so later on a similar working.

Plate 231,232: These two views show the same train (London-Perth) at the same location but from opposite sides of the line just north of Carlisle. The locomotive is *Duchess of Norfolk* and only the date is different, the upper view being August 1962 and the lower one a few weeks earlier.

S.C. Crook, Derek Cross

Plate 233 (above): The down Perth again in August 1962 — this time in Floriston Woods, north of Carlisle, *City of Liverpool* in charge.

S.C. Crook

Plate 234 (below): Beattock station and the starting point of another tough climb. In this view, at least two freights wait their turn while *City of London* gets the road with a northbound express.

W.J.V. Anderson

Plate 235 (above): For my money, this has to be one of the finest Duchess pictures ever taken on Beattock bank and I make no apologies for publishing it again. It shows a northbound Glasgow sleeping car train near Greskine on a July morning in 1959 behind *Princess Alice*. It was only in the morning that the sun was on the right side of the line to enable pictures to be taken without interference from telegraph poles on this stretch of line.

W.J.V. Anderson

Plate 236, 237 (opposite): When the sun moved to the west side of the line, photography always became more difficult on Beattock bank because of the wretched telegraph poles. Nevertheless, many fine pictures were taken, including this pair by Eric Treacy. They show *Queen Mary* and *City of Leicester* respectively during the mid-1950s.

Plate 238, 239: The 'Royal Scot' stops for water at Beattock summit. It was rare for the crack expresses to make unscheduled water stops and one can only presume that on this occasion, a set of water troughs were either out of action or empty. Fortunately, the photographer was on hand when Duchess of Buccleuch made this out of course halt.

Eric Treacy

Plate 240, 241: Like Shap, the southbound ascent to Beattock was less spectacular but probably more demanding. On this page, the southbound climb is shown in contrasting moods. *Above,* Queen Mary heads into the night with the up West Coast Postal at Elvanfoot while the *lower* picture shows a late morning scene with City of Manchester in charge of the up 'Royal Scot'. The smoke deflectors are clearly doing their work on this occasion.
  David Anderson, W.J.V. Anderson

Plate 242, 243: Further scenes on the southbound climb to Beattock, this time near Crawford. In the *upper view*, City of Lancaster heads the up 'Mid-day Scot' while *below*, Duchess of Buccleuch is in charge of the southbound 'Royal Scot'. Both pictures were taken in 1957.

Eric Treacy, W.J.V. Anderson

Plate 244-6: Glasgow Central — journey's end, or beginning. In the *upper view*, recently de-streamlined *City of Glasgow* herself leaves Central station with a southbound express in 1947. In the *centre view*, the famous roof makes a nice background to *City of Chester* in early BR days while the *last picture* is a brief reminder of the day in 1937 when the first blue streamliner to cross the border was shown off in Glasgow — *Queen Elizabeth*.
    BR LMR, Gavin Wilson collection

DUCHESSES IN DECLINE

Like all good things, the reign of the Duchesses had to end but there were those of us (including, let it be said, not a few railwaymen) who thought that their wholesale withdrawal in 1964 might just be a year or two premature. Certainly, some of the summer extra workings in 1965/6 seemed to be in want of Class 8 power — but such was to be the course of history. In their final years, the Duchesses, like others, had to suffer relegation to freight, parcels and minor duties, extended schedules (because of electrification) which hardly taxed their power and other lesser indignities. Nothing, however, could spoil their essential dignity, even though it was pitiful to see them on three coach locals. Occasionally, of course, they deputised to good effect for diesel failures and they were often called upon to operate in wildly unfamiliar territory on a variety of rail tours during the early 1960s. In fact, the last time the writer saw one in service was on just such a rail tour — *City of Hereford* going magnificently over the Settle-Carlisle main line in July 1964. They all vanished during the autumn and we felt sad.

Plate 247 (left): It was not until after 1960 that the reign of the Duchesses on the West Coast began to end. By 1963 most expresses were diesel hauled and *King George VI* was probably deputising for a diesel at Beattock with a lightweight train.
Derek Cross

Plate 248: When released from some of their former main line duties, Duchesses were relegated to lesser tasks, an interesting example of which was the down 'Lakes Express', a regular Duchess turn in the early 1960s. *City of Bradford* is seen with this train on Shap during July 1961 — Keswick portion only, of course.
Derek Cross

Plate 249-51: Local passenger and excursion duties fell to the lot of several Duchesses in former GSWR territory. At the *top of the page*, *Duchess of Buccleuch* is at Crosshouse with a Kilmarnock-Glasgow train in June 1962. The first three coaches are also interesting — all non-corridor brakes of LMS, LNER and BR origin respectively.

Falkland Junction is the venue of the *centre picture* of an Ayr-Glasgow race special in July 1963 with *Princess Alice* in charge. In spite of the express headcode, the train is composed of high capacity suburban stock.

The *final view*, taken at Dumfries in June 1963 shows what might be termed a prototype version of a boxed toy train set — *Princess Alexandra* doing its best to maintain a vestige of dignity at the front of a four coach Carlisle-Glasgow stopping train. *Derek Cross*

Plate 252, 253: *City of Nottingham* was one of the last Duchesses to be scrapped and was a popular choice for excursions. *Above*, the locomotive hurries along with a fast freight at Winwick while at the *left*, she is in charge of the RCTS 'East Midlander' rail tour passing Appleford Halt on 9th May 1964.

J.R. Carter, A.H. Molyneaux

Plate 254, 255: *City of London* was another late survivor, also called upon for excursions. The two pictures show the locomotive backing onto its train at Kings Cross (*right*) and at Doncaster (*below*) on the occasion of a special working in 1963.

Gavin Wilson collection

Plate 256 (above): Duchess of Hamilton waits with a parcels train in the loop at Penrith to allow a diesel hauled express to overtake hauled, coincidentally by No. D229! No. 46229 was, however, destined to have the last laugh for she is now preserved for all time at the National Railway Museum.

Derek Cross

Plate 257 (below): One of the last regular Duchess duties was the 06.00 Carlisle-Perth parcels train and *City of Hereford* was in charge of this working when seen passing Greskine box in June 1964.

Derek Cross

Plate 258,259: Super-powered milk! In the *upper view*, *City of Bradford* is in charge of a down milk train at Hest Bank in August 1961 while *below*, *City of London* climbs the northern side of Shap with a similar southbound working on Good Friday, 1963.

Gavin Wilson, S.C. Crook

Plate 260, 261: Parcels workings could, at least usually, provide a load worthy of a Duchess and in these pictures, taken in the summer of 1963, *City of Glasgow* heads north over Dillicar troughs *(above)* and *Duchess of Gloucester* passes Peat Lane between Oxenholme and Tebay *(below)*.

Derek Cross

Plate 262: Duchess of Gloucester at Penrith on 22nd August 1964, waiting to take charge of the up 'Lakes Express'.
Derek Cross

Plate 263: The same engine, two months earlier at Euxton Junction with a parcels train.
Photomatic

Plate 264: The ultimate indignity — City of Lichfield leaving Carlisle with but one van and a brake in April 1964.
S.C. Crook

Plate 265, 266: There were times, however, when even in their twilight years, the Duchesses could relive their great days — two such occasions are illustrated here. In the *upper view*, City of Stoke-on-Trent climbs to Shap past Thrimby with the up 'Royal Scot' on 8th September 1963 while *below*, City of Lichfield is on the same duty at Wreay in December the same year. In both cases the Pacifics were replacing failed diesels and in the case of the earlier view it was almost certainly the last time that the 'Royal Scot' headboard was carried by a steam locomotive.

S.C. Crook

Plate 267, 268: Duchess farewell. No. 46256 *Sir William A. Stanier, FRS* was the last survivor of all and is seen *above*, only a few weeks before scrapping, on an up August Bank Holiday extra leaving Carlisle in 1964. She was scrapped in October and the following months saw a melancholy procession of Duchesses en route to the scrap yards — such as *Duchess of Norfolk* (*below*) at Troon harbour in January 1965.

S.C. Crook, Derek Cross

DUCHESSES IN ASPIC

The somewhat precipitate withdrawal of the engines in late 1964, being slightly ahead of the later vogue for wholesale private preservation, made it seem for a time as though only the officially preserved No. 46235 *City of Birmingham* would survive for posterity. Then, mirabile dictu, two more were saved by the Butlin organisation — *Duchess of Sutherland* and *Duchess of Hamilton*. These two, somewhat incorrectly repainted, were located at Ayr and Minehead for almost ten years until Butlin's Ltd generously agreed to their recovery and further restoration — one to Bressingham and one to the National Railway Museum, York. *Duchess of Sutherland* is now beautifully restored to original LMS livery at Bressingham and it was a happy choice which sent the transatlantic wanderer No. 46229 to a final resting place at York, whose officially preserved No. 46235 was, by then, comprehensively walled in at the Birmingham museum!

One can, perhaps regret that No. 6220 was not saved or, even more, that *Sir William A. Stanier, FRS* did not survive. One can also look with envy on the no fewer than six A4s which still exist (albeit two across the Atlantic). But never mind, there are still three Duchesses for posterity to admire — all carrying different colour schemes — and, at the time of writing, arrangements are well in hand for the NRM example *Duchess of Hamilton* to be restored to working order again — probably by the time this book appears. With the Settle-Carlisle main line recently approved for steam working once more, maybe we can again look forward to hearing that low pitched hoot and dark four beat exhaust echoing back from the northern fells as in more halcyon days. Ais Gill may not be Shap or Beattock — but it's not a bad substitute!

Plate 269, 270: City of Birmingham *at work in 1962 at Willesden Junction* (above) *and being manoeuvred into position at the Birmingham Science Museum* (below) *in 1966.*

Photomatic, John Edgington

Plate 271 (below): Duchess of Sutherland *went on show at Heads of Ayr Holiday camp for many years. Here she is seen en route for Bressingham, passing Mossblown Junction, Ayrshire on 1st March 1971, still wearing the incorrect version of the pre-war LMS livery.*

Derek Cross

Plate 272-5 (above): Restoration work at Bressingham 1972-4. In the upper views, the last tubes are withdrawn from the boiler prior to overhaul in October 1972. Less than two years later (July 1974) the engine, now correctly painted, receives a final polishing up prior to steaming through the woods at this well known centre.

John Edgington

Plate 276: Shades of things to come? *Duchess of Hamilton* at speed with the 'Caledonian' — Hest Bank 1957.

D. Jenkinson

FINAL APPRAISAL

There never was any such thing as a 'best ever' steam locomotive. Some were good, some bad, some indifferent. Some were ideally suited for one particular task and woefully inadequate for others. Some designs seemed to do anything asked of them and more besides while others never quite seemed to be as good as they should be. Such was the fascination of steam — and still is for those of us who like to debate the subject. Since it is all over now, why should we worry? I think it is bound up with a sense of history. There *were* memorable locomotives both in Britain and overseas which deserve their place on the canvas of world steam development and the man who pays no heed to the events which helped to form his own nation's history — whether in engineering or art is less of a man for his omission.

Do the Duchesses merit inclusion in this august company? Well, they were Britain's most powerful express steam engines — this much is fact, just as it is fact that the LNER A4 was the fastest steam locomotive ever built. But were the Duchesses more than this? I think they were. There may have been better looking latter day engines (this is a subjective point anyway) and there may have been more glamorous or thermodynamically efficient ones but taken all round I cannot think there was ever a better British express steam engine in relation to the era when it was born and the environment in which it had to work than the Stanier Duchess. In sheer cost-effective terms (a modern enough phrase but one which, perhaps, more than any other British railway the LMS always had well to the fore) the Duchesses were a good buy. No new boilers were ever built for them after 1948 and only four spares were needed for the whole class. They generally ran almost 100,000 miles between heavy repairs and at some 67,000 miles per year, their average annual revenue mileage was well above that of the competition. They were 'first timers' from the moment that 6220 turned its wheels under power in 1937 and, cosmetic variations set aside, the only major change was to fit a double chimney. They were easy to drive — provided the driver remembered how much steam was still left in the system after he had closed the regulator; and once the firemen had mastered the large grate — which generally did not take long to learn — they never lacked for steam. Full regulator and 25% cut off meant real business with a Duchess; yet even at high power outputs, they were not extravagant on coal. Above all, their designer liked them and their crews loved them. For such an imprecise machine as a steam locomotive, that probably says it all.

Plate 277: Last of the line — *City of Salford* rouses the echoes at Greskine with a relief sleeping car train in July 1964.

Derek Cross

THE DUCHESSES IN SUMMARY — PRINCIPAL CHANGES

Number	Name	To Traffic	Double Chimney	Smoke¹ Deflectors	BR Numbers	Smoke box to Normal	LMS Streamline Blue	LMS Streamline Red	LMS Streamline Black	Principal Livery Changes² Red	LMS Non-streamline Wartime	1946³	BR Experimental⁷ Blue	BR Experimental⁷ Black	Blue¹⁰	Green¹¹	BR Standard¹⁰ Green¹²	BR Standard¹⁰ Red¹³	Red¹⁴	Mileage²⁰	Withdrawn
6220	Coronation	6/37	12/44	9/46	7/48	12/55	6/37		3/44			10/46			1/50	8/55	6/58			1,321,682(59)	4/63
6221	Queen Elizabeth	6/37	11/40	5/46	10/48	9/52	6/37	3/41	8/44			7/46			10/50	1/53	11/57			1,308,644(59)	5/63
6222	Queen Mary	6/37	8/43	5/46	9/48	8/53	6/37		10/44			5/46			11/50	12/52	1/59			1,420,944(62)	10/63
6223	Princess Alice	7/37	11/41	8/46	3/49	8/55	7/37		2/44			8/46			4/50	10/52	11/58			1,433,672(62)	10/63
6224	Princess Alexandra	7/37	5/40	5/46	5/48	10/54	7/37		10/44			7/46	5/48		6/51	6/53	8/57			1,430,317(62)	10/63
6225	Duchess of Gloucester	5/38	6/43	2/47	6/48	1/55		5/38	4/44			2/47			4/50	3/56		8/58	10/61	1,742,624	9/64
6226	Duchess of Norfolk	5/38	7/42	6/47	9/48	11/55		5/38	1944			6/47		11/48	5/49	9/56		11/58	N.K.	1,456,947(59)	9/64
6227	Duchess of Devonshire	6/38	12/40	2/47	5/48	5/53		6/38	1/44			2/47	5/48*		7/50	5/53	2/58			1,412,644	12/62
6228	Duchess of Rutland	6/38	9/40	7/47	7/48	1/57		6/38	1944			7/47			1950	8/55			6/58	1,394,049(58)	9/64
6229	Duchess of Hamilton	9/38	4/43	11/47	7/48	2/57		9/38	1945			12/47			5/50	4/53		9/58	1962³	1,533,846²¹	2/64
6230	Duchess of Buccleuch	6/38	10/40	9/46	5/48					6/38		*				5/52	8/58			1,464,238(62)	11/63
6231	Duchess of Atholl	6/38	6/40	9/46	5/48					6/38	1945	*			N.K.	9/55	5/58			1,472,439	12/62
6232	Duchess of Montrose	7/38	1/43	2/45	5/48					7/38		*				11/51¹⁷	3/59			1,420,948	12/62
6233	Duchess of Sutherland	7/38	3/41	9/46	10/48					7/38¹⁸	Never	1947			8/50	5/57	2/61			1,644,271(62)	2/64
6234	Duchess of Abercorn	8/38	2/39	3/46	10/48					8/38		*⁵				8/56	8/58			1,494,604(59)	1/63
6235	City of Birmingham	7/39	New	4/46	5/48	7/52		7/39	3/43		*⁶	4/46			1950	4/53	3/58¹⁹			1,566,677(63)	9/64
6236	City of Bradford	7/39	New	12/47	4/48	11/53		7/39	4/44			12/47			Never	5/52		7/58	11/59	1,629,412(63)	3/64
6237	City of Bristol	8/39	New	1/47	7/48	5/56		8/39	1945			1/47			5/49	4/55	10/57			1,477,715(59)	9/64
6238	City of Carlisle	9/39	New	11/46	3/49	10/53		9/39	1943			11/46		1949		4/52			6/58	1,602,628(63)	9/64
6239	City of Chester	9/39	New	6/47	8/48	2/57		9/39	1943			6/47			10/50¹⁵	4/56	3/58			1,544,194(59)	9/64
6240	City of Coventry	3/40	New	6/47	6/48	5/57		3/40	11/45			6/47			1/50¹⁶	5/52		7/58	4/62	1,685,042(63)	9/64
6241	City of Edinburgh	4/40	New	1/47	5/48	2/58		4/40	1945			1/47	5/48		10/49	9/53	2/58			1,425,987(59)	9/64
6242	City of Glasgow	5/40	New	3/47	6/48	11/53		5/40	1944			3/47			8/49	11/53	1961		9/59	1,555,280(62)	10/63
6243	City of Lancaster	6/40	New	5/49	4/48	11/58		6/40	1/44			Never			6/49	1953	11/57	11/58		1,526,292(63)	9/64
6244	King George VI⁷	7/40	New	8/47	8/48	7/53		7/40	1/44			8/47	9/48*		1/49	5/53		10/58	8/60	1,394,153(59)	9/64
6245	City of London	6/43	New	8/47	8/48	12/57			6/43			8/47			Never	4/53			12/57¹⁴	1,408,315(63)	9/64
6246	City of Manchester	8/43	New	9/46	11/48	5/60			8/43			9/46		2/49	Never	6/54	10/58		5/60	1,168,596(59)	1/63
6247	City of Liverpool	9/43	New	5/47	11/48	5/58			9/43			5/47			Never	5/53	5/58		1959	1,388,187	5/63
6248	City of Leeds	10/43	New	12/46	3/49	6/58			10/43			12/46		3/49	N.K.	6/54	6/58		1962	1,136,509(59)	9/64
6249	City of Sheffield	4/44	New	11/46	4/48						4/44	11/47			N.K.¹⁵	7/56				1,098,157(62)	11/63
6250	City of Lichfield	5/44	New	3/46	2/49						5/44	10/47			5/50	2/53	11/57			1,353,526(63)	9/64
6251	City of Nottingham	6/44	New	8/46	5/48						6/44	8/47		5/49	Never	10/55		11/58	1960	1,236,546(63)	9/64
6252	City of Leicester	6/44	New	3/45	6/48						6/44	1947		4/49		2/52	2/59			1,231,032	5/63
6253	City of St. Albans	9/46	New	New	9/49							9/46			Never	3/54	1/58			932,417(59)	1/63
6254	City of Stoke-on-Trent	9/46	New	New	7/49							9/46			10/51¹⁵	9/55		9/58	5/60	1,103,041(63)	9/64
6255	City of Hereford	10/46	New	New	6/49							10/46			8/50	4/53	11/57			836,858(59)	9/64
6256	Sir William A. Stanier, FRS	12/47	New	New	5/48							12/47		10/48	5/52¹⁵	6/54		5/58	9/59	1,016,060(63)	10/64
46257	City of Salford	5/48	New	New	New							5/48				4/53	1/58			806,758(59)	9/64

NOTES

¹ Also date of de-streamlining where applicable.
² Many of the dates quoted are based on observation and, apart from the first livery carried, do not necessarily denote the precise date of which the change style appeared.
³ In general this was the fully lined version — see Plate 84. Some early repaints had less lining and non standard insignia including (at least) 6222, 6223, 6224, 6235.
⁴ The first five non streamliners were still red in 8/44. Only 6231 is known to have received wartime black and 6233 was still red when fitted with smoke deflectors. All except 6234 below — note 5) probably received the post-war 1946 livery.
⁵ Painted experimental blue/grey (3/46) until (at least) 1/48. Prototype scheme for post-war LMS livery but not adopted.
⁶ Immediately after de-streamlining ran for a short period with wartime insignia, no lining.

⁷ Generally with LNWR style lining.
* Possibly repainted LNWR style black later — not confirmed.
⁸ This was probably the prototype for the standard BR blue. Colour was paler blue and lining was black/yellow, later changed to black/white.
⁹ Only confirmed examples are quoted, there may have been others, especially BR Blue. Letters N.K. indicate livery carried but exact dates not known.
¹⁰ Mileage figures were not always recorded to scrapping dates. Figure in brackets indicates the year to the end of which the figures apply, e.g. 6220's mileage of 1,321,682 applies to the end of 1959. Where no date is given, the mileage is as at withdrawal.
¹¹ With original tender emblem) These are dates observed in traffic — All locomotives carried BR green
¹² With later tender emblem) at some time but whether all engines carried both emblems is not known.
¹³ With BR style lining set in from panels — all repainted eventually with LMS pattern lining — exact dates not always available but the style was short-lived.
¹⁴ LMS style lining at panel edges. Only 16 locomotives were repainted red and all were based on the London Midland Region. Dates are generally as observed in traffic.

¹⁵ Ran until at least 4/53 in this style.
¹⁶ Still blue in 6/54 but believed to have been green before repainting red in 1958.
¹⁷ First to be painted BR green.
¹⁸ First to be painted BR red — and this locomotive always had LMS type lining with the red livery.
¹⁹ Livery in which preserved example is restored.
²⁰ Mileage excludes period 1939-42 (in USA).
²¹ Named *City of Leeds* until 4/41.